Adult Numeracy Development

Theory, Research, Practice

Volumes in the series include:

Literacy Among African-American Youth
(Vivian L. Gadsden and Daniel A. Wagner, eds.)

Adult Numeracy Development: Theory, Research, Practice
(Iddo Gal, ed.)

What Makes Workers Learn
(Donald Hirsch and Daniel A. Wagner, eds.)

International Perspectives on the School-to-Work Transition
(David Stern and Daniel A. Wagner, eds.)

Adult Basic Skills: Innovations in Measurement and Policy Analysis
(Albert Tuijnman, Irwin Kirsch, and Daniel A. Wagner, eds.)

The Future of Literacy in a Changing World: Revised Edition
(Daniel A. Wagner, ed.)

Forthcoming

Adult Literacy Research and Development Vol. 1: Learning and Instruction
(Daniel A. Wagner, ed.)

Adult Literacy Research and Development Vol. 2: Programs and Policies
(Daniel A. Wagner, ed.)

Adult Numeracy Development

Theory, Research, Practice

edited by

Iddo Gal
University of Haifa

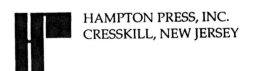

HAMPTON PRESS, INC.
CRESSKILL, NEW JERSEY

Library of Congress Cataloging-in-Publication Data

Adult numeracy development : theory, research, practice / edited by Iddo Gal
       p. cm. -- (Literacy)
    Includes bibliographic references and index.
    ISBN 1-57273-232-6 (cl) -- ISBN 1-57273-233-4 (ppb)
      1. Mathematics--Study and teaching. 2. Adult education. I. Gal,
    Iddo. II. Series
QA11.D43 2000
510'.71--dc21

00-028114

Hampton Press, Inc.
23 Broadway
Cresskill, NJ 07626

Contents

Series Preface

When people think of problems of literacy and illiteracy in today's world, there is a strong tendency to focus primarily on the elements of reading and writing. Yet, virtually all of the myriad definitions of literacy utilized by the United Nations over the last half-century include calculating or mathematics as part of the definition of literacy. For historical reasons, as discussed in this important volume, numeracy—especially adult numeracy—has not received that kind of attention to date that is needed in order to create a more fully literate and numerate world. With the rapidly changing economies in all parts of the globe, there is little question that all adults, whether in rich countries or poor countries, will need to deploy greater facility with numeracy skills as part of an increasing trend toward lifelong education.

This volume, *Adult Numeracy Development: Theory, Research, Practice*, begins to open up key issues that will help a variety of educational specialists reconsider the nature of numeracy, and how to promote it in a range of formal and non-formal settings. Dr. Iddo Gal, the editor, has brought together a diverse set of authors from around the world, all of whom are working to explore various parts of the growing field of adult numeracy and mathematics education for adults.

The present series, Literacy: Research, Policy and Practice, is one attempt to break down the walls that partition literacy into such separate intellectual territories, or that focus only on the narrow processes of reading acquisition. In this series, we try to find interconnections not only among the three major segments of literacy and numeracy specialists as denoted in the series subtitle, but also across the life span (children and adults) and across ethnic, linguistic and cultural groups. This series, we hope, will provide an opportunity to make connections across various knowledge bases and expand the possibilities not only to achieve a better understanding of literacy in the past, present, and future, but also to lead to a future which is a more literate and numerate place to live.

Daniel A. Wagner
International Literacy Institute
National Center on Adult Literacy
University of Pennsylvania

Preface

The field of adult numeracy is a growing area of practice and research in many countries. Numeracy has a recognized role in contributing to the empowerment, effective functioning, economic status, and well being of citizens and their communities. Yet, relatively few comprehensive publications have so far been addressed at professionals interested in numeracy development. This book was designed as a resource for educators, trainers, curriculum developers, program managers, and researchers interested in adult education, literacy education, workplace training, as well as in mathematics education in diverse learning contexts.

Early planning of this book started a few years ago when I served as Director of the Numeracy Project at the National Center on Adult Literacy (NCAL), a federally-supported R&D center at the Graduate School of Education of the University of Pennsylvania, Philadelphia. It was clear that the emerging field of numeracy education could benefit from new high-quality and conceptually sound resources, yet that the diversity of learners, teachers, programs, and instructional challenges requires the coordination of contributions of diverse experts to produce a comprehensive and useful publication.

In developing this volume, chapters were sought from authors with background in mathematics education, adult education, assessment, educational research, teacher training, curriculum development, language instruction, workplace learning, and work with populations with special learning difficulties. We sought colleagues able to address both theoretical aspects and classroom realities and to reflect on the challenges and dilemmas involved in implementing new as well as proven teaching methods when working with diverse types of adult students. While most of the authors are based in the United States, colleagues from the United Kingdom, The Netherlands, Israel, Australia, Canada, and Malaysia have also contributed to this volume, and their chapters help make this volume of interest to international audiences.

A special effort was made during the planning stages to make sure that chapters solicited for this book are organized, written, and interrelated so that the final manuscript presents readers with a broad yet cohesive set of ideas and suggestions. Chapters were reviewed by a panel of experienced adult and mathematics educators to evaluate the degree to which their language is accessible and their messages clearly explained in light of the known diversity of learners, practitioners, and learning contexts. The editorial process also ensured that all chapters open with a brief statement that lists the key questions the chapter addresses, and end with a summary section discussing implications and reiterating key ideas raised. Readers can examine these opening and closing sections for a quick determination of each chapter's contents.

A project of this magnitude would have been impossible without the support of numerous individuals and organizations.

I would like to thank the National Center on Adult Literacy (NCAL) at the University of Pennsylvania, and its director, Daniel Wagner, for supporting the early stages of my work on this book. Partial funding was provided through grants from the U.S. Department of Education, and also from the Departments of Labor and Health and Human Services.

Several members of NCAL's staff contributed to this book: I am grateful to Caroline Brayer-Ebby and Lynda Ginsburg for their role in reviewing drafts and editing certain chapters. Special thanks and gratitude go to Ashley Del Bianco-Stoudt, who served as the project's Editorial Coordinator. Ashley managed the extensive review process, compiled reviewers' comments, skillfully contributed to editorial work on several chapters, and made many useful suggestions.

The University of Haifa has supported my work and the inordinate amount of copying, faxing and e-mail traffic required to complete a book of this kind.

Numerous professionals involved in adult and mathematics education in various organizations and academic institutions volunteered to participate in the review process. I would like to acknowledge their generous investment: Leslie Arriola, Dianne Arvizu, Sandi Braga, Rose

Brandt, Sandra Choukroun, Joyce Claar, Rebecca Clark Jordan, Betty Conaway, Richard Cooper, Donna Curry, Elizabeth van Dusen, Harriet Hartman, Ron Kindig, Jereann King, Ellen McDevitt, Peggy McGuire, Melissa Mellissinos, Rolf Parsons, Laura Roberts, Annette Sanger, Mary Ann Shope, Sally Spencer, Jean Stephens, Carol Thornton.

I would like to thank our many contributors, who remained focused on the goal of sharing their insights, findings, observations, and experiences, while enduring multiple review cycles and editing demands needed to fit their contributions into an integrated volume.

Throughout the work on this book, I have been helped by the friendly advice and suggestions from Barbara Bernstein, President of Hampton Press, regarding both content and technical matters.

Lastly, to my wife Aviva, and to my daughters Naama and Danna, many thanks for bearing with the many months it took to complete this project. I also thank you for being on the lookout, whenever you were reading the newspaper, shopping, or watching TV, for the varied instances and uses of numbers, averages, probabilistic statements, and other manifestations of numeracy tasks in our everyday life.

The chapters in this volume by no means exhaust all issues encountered in environments where numeracy skills are being taught or learned, formally or informally. Yet, taken as a whole, the chapters constitute a rich resource that for the first time offers, within a single volume, a coherent set of frameworks and approaches to developing and studying numeracy-related knowledge, thinking processes, and dispositions. My hope is that this volume will stimulate a scholarly discourse within the professional community involved in numeracy, mathematics, and literacy education, and contribute to the recognition both by academic institutions and policy-makers that numeracy is an important and unique area of research and practice.

Iddo Gal

Contributors

Leslie Arriola has been researching and teaching adult numeracy for over twelve years, and was a member of the Massachusetts Adult Basic Education Math Team. Her interests include basic mathematics problem solving and misconceptions, and effective teaching practices. She has worked with teachers and learners in schools, community colleges, universities, and adult learning centers. Presently she consults to schools on implementing a hands-on/inquiry-based science curriculum.

Dorothy Cebula is Director of Disability Resources and Services at Temple University. She has extensive experience with disability services both at Temple University and at a large community college in New Jersey, and has certification as a Learning Disabilities Teacher Consultant. Her teaching experience has ranged from elementary through the graduate level while her research focus has been on adults with histories of learning difficulties.

Robert Clemen is a Professor of Decision Sciences at the Fuqua School of Business, Duke University. He has participated actively over the years in developing curriculum materials to help people of all ages learn to avoid pitfalls and improve decision-making processes. He authored *Making Hard Decisions: An Introduction to Decision Analysis* (1991), and co-authored *Creative Decisionmaking: A Curriculum Materials Guide for Secondary-School Educators* (1996).

Diana Coben is a Senior Lecturer in the School of Continuing Education, University of Nottingham, England. She is a founder and member of Adults Learning Mathematics (ALM), an international research forum. She is co-editor of *Perspectives on Adults Learning Mathematics: Research and Practice* (2000), and co-author of *Carefree Calculations for Healthcare Students* (1996) and *The Numeracy Pack* (1984).

Joy Cumming is an Associate Professor in the Faculty of Education at Griffith University, Queensland, Australia, and Head of the School of Cognition, Language and Special Education. She has been involved in national and international research on literacy and numeracy for all ages, with an emphasis on learning and assessment. A major focus of her current work are contextualized and performance assessment, particularly for work-related training, and the transfer of learning across contexts.

Donna Curry is staff development specialist for the Center for Adult Learning and Literacy at the University of Maine, Orono. She is responsible for providing leadership and guidance to adult educators at a statewide level. Since 1995, she has been an active partner on the National Institute for Literacy Development Team crafting adult education standards, in particular "use math to solve problems and communicate."

Jim Foerch taught adult learners of diverse ages and backgrounds and adults with special needs since 1975 in the Grand Rapids, Michigan, Community Education program. His focus was always on assessing students' skills, identifying realistic goals and designing learning activities to help them reach those goals. Since 1991 he has been teaching math and science to teens in a Grand Rapids alternative high school.

Iddo Gal teaches in the Department of Human Services, University of Haifa, Israel. With a background in cognitive and applied psychology, he is interested in adult numeracy, statistical reasoning, and empowerment processes. He co-edited *The Assessment Challenge in Statistics Education* (1997). Formerly, he directed the Numeracy Project at the National Center on Adult Literacy at the University of Pennsylvania, and led NSF-funded projects on statistics and mathematics education.

Lynda Ginsburg is a Senior Researcher and Project Director at the National Center on Adult Literacy at the University of Pennsylvania. She has extensive experience teaching mathematics learners in high school, adult education, community college, and workplace settings. She has also developed and taught courses and workshops for K-12 teachers and adult educators on enriching instructional practice. She is particularly interested in adults' acquisition of numeracy and technology skills.

Robin Gregory is a Senior Researcher with Decision Research in Vancouver, Canada and Associate Director of the Eco-Risk Research Centre at the University of British Columbia. He works on applied and theoretical problems in decision making, with a focus on environmental and health risks. He works with and trains college-level and secondary-school teachers to develop classroom materials on decision making that clarify how understanding values and probabilities can lead to better choices.

Mieke van Groenestijn teaches at the Utrecht University of Professional Education in the Netherlands, where she leads post-graduate courses for teachers in adult education and special education. She developed assessment tools and a series of learning books on functional mathematics for adults in adult basic education. She is a member of the international team developing numeracy assessment for the International Life Skills Survey.

Mary Harris has recently retired from the University of London's Institute of Education, but continues her work on mathematics in everyday life through consultancies. With a Millennium Fellowship she is now involved in a project aiming to identify and accredit the mathematics implicit in patchwork and quilting courses taught by the National Federation of Women's Institutes in England and Wales.

Harriet Hartman is an Assistant Professor of Sociology at Rowan University in Glassboro, New Jersey. Her main fields of teaching and research are sociology of education, socialization, family, and gender in comparative perspective. She is the co-author of *Gender Inequality and American Jews*.

Peter Kloosterman is a professor of mathematics education at Indiana University and chair of the Department of Curriculum and Instruction. Although his specialization is student motivation to learn mathematics, he is also involved in several projects that promote teaching mathematics through applications in K-12 and adult settings. A listing of some of his publications and more detail about his teaching and service activities can be found on his home page (http://www.indiana.edu/~pwkwww).

Betty Hurley Lawrence is a professor at SUNY Empire State College, which focuses primarily on the educational needs of adult learners. She currently works with the Center for Distance Learning and the Office of International Programs. She has a doctorate in mathematics education from the University of Rochester. Her work focuses on facilitating the learning of mathematics and on using technology to facilitate that learning, including the use of web-based courses she developed.

Wim Matthijsse works as consultant at CINOP, Den Bosch, the Netherlands. CINOP is a national centre for the innovation of education and training in the Netherlands, and specializes in professional training and adult education. Matthijsse specialty is in numeracy and mathematics education. Currently he works on the development of a multimedia program for learning mathematics, as part of a remedial mathematics program for individuals starting vocational training in regional learning centers.

Karen Hicks McCormick is an Associate Professor at Southeastern Louisiana University, where she teaches preservice and inservice teachers. Her specialty areas include integrating reading and writing across the curriculum and methods for facilitating effective teaching behaviors and maximum student interaction.

Bin Hassan Mohamad-Ali obtained his doctorate in mathematics education from Indiana University. His areas of research and interest include students' attitudes toward mathematics, remedial mathematics programs, and use of technology in teaching mathematics. Currently he is a professor at the MARA University of Technology in Malaysia, where he also teaches actuarial mathematics and risk theory.

Martha Sacks is an Assistant Professor in the Department of Special Education and Individualized Services at Kean University, New Jersey, and also the Coordinator of the Undergraduate Program in her department. She has experience teaching in remedial and special education programs from elementary through the graduate level. Her research has focused on remedial strategies in writing and mathematics as well as study strategies for adults with attention deficit disorder.

Elizabeth Wadlington is an Associate Professor at Southeastern Louisiana University. She teaches literacy courses and is especially interested in the problems of adults with dyslexia. In the past, she taught math education and directed an adult literacy program. She continues to conference with adults about their learning problems.

Lynda R. Wiest taught elementary and middle school students for eleven years in Pennsylvania. She attained a Ph.D. in Curriculum Studies/Mathematics Education at Indiana University, and presently is an Assistant Professor of Elementary Education at the University of Nevada, Reno. Her professional interests include mathematics education, educational equity, and teacher education.

Introduction

Iddo Gal
University of Haifa

Many thousands of adult educators are involved on a daily basis in math instruction in both developed and developing countries all over the world. Yet, most of these teachers have had little opportunity to gain the training needed to meet the math-related goals and needs of adult learners, as well as to satisfy the increasing demand by public agencies, community organizations, or business organizations to improve adults' numeracy or mathematical literacy. Administrators and agencies involved in adult education often lack a clear understanding of what numerate behavior entails and the nature of the many challenges involved in developing numeracy-related skills and dispositions of adult learners. As a result, adult education programs too often are unable to fully reach their goals of ameliorating inadequate numeracy and related literacy skills, imparting new skills that learners can effectively transfer and apply in new diverse contexts, and helping learners become informed and empowered consumers, parents, workers, and citizens.

This volume aims to invigorate the field of adult numeracy education by serving as a resource for teachers, trainers, and curriculum developers involved in math teaching in adult or literacy education in diverse contexts. In addition, the chapters in this volume are designed to serve as background readings for academic courses focused on preparing the next generation of adult numeracy practitioners and program planners, and to highlight several key issues that could be of interest to researchers interested in adults' mathematical thinking.

The 16 chapters in this volume introduce recent views about the nature of numeracy, discuss instructional principles, recommend teaching practices tailored to adult needs, examine assessment strategies, and review relevant research findings that can support work by practitioners and program planners.

Section I, Perspectives on Numeracy, includes four chapters that present broad conceptions of the goals of adult numeracy education, outline needed skills and dispositions, and review frameworks and principles that should guide or support the teaching-learning process.

In Chapter 1, Iddo Gal examines issues and tensions involved in defining the goals of adult numeracy education. In discussing the relative and dynamic nature of numeracy skills, the existence of multiple perspectives on numeracy, and the theoretical and functional needs to create and strengthen links between literacy and numeracy skills and their acquisition, Gal raises implications for instruction and professional development and presents challenges that some of the other chapters in the volume begin to address.

Diana Coben examines research findings about numeracy and adult learning in Chapter 2. From an international perspective, Coben examines what we know about how adults learn (in general) and how they develop numerical concepts and mathematical skills, drawing connections between research findings and their potential influence on instructional practices in numeracy education.

In Chapter 3, Peter Kloosterman, Ali Hassan, and Lynda Wiest relate the curriculum and teaching standards of the National Council of Teachers of Mathematics to adult numeracy education. Suggestions in line with NCTM recommendations are made for designing a learning environment for adults in which mathematical problem solving is taught through effective questioning, collaborative learning, and appropriate use of instructional resources.

In Chapter 4, Robert Clemen and Robin Gregory assert that being an active and contributing member of society requires individuals to make informed, valid decisions. This chapter discusses the kinds of skills that adults need to make difficult decisions, including structuring a decision situation, understanding uncertainty, and identifying crucial tradeoffs; the authors also suggest some guidelines for integrating decision-making instruction into the literacy and numeracy curriculum, and thus help to broaden the scope of what may (and should) be included in numeracy education.

The chapters in *Section II, Approaches to Instruction,* examine critical issues encountered by adult numeracy educators as they plan and implement mathematics learning experiences for diverse groups of learners.

In Chapter 5, which builds in part on the foundations and visions laid out by the first four chapters in Section I, Lynda Ginsburg and Iddo Gal address the concerns of the many adult educators who want to improve the ways they teach mathematics, but who are uncertain about how to begin this process of change. Their chapter sets forth 13 principles that should guide adult numeracy education and suggests specific instructional practices and strategies to guide implementation and improve the teaching-learning process.

Drawing on his experiences as an adult mathematics educator, in Chapter 6 James Foerch describes the parameters of diversity and the characteristics commonly found among adult learners. Foerch suggests how to identify learners' unique motivations, needs, and strengths to assist students in setting realistic, achievable goals for numeracy learning.

In Chapter 7, Dutch author Wim Matthijsse presents an innovative framework developed in the Netherlands for teaching addition and subtraction in a basic-level numeracy course for adults. This method of instruction, based on the "Realistic Mathematics" approach, aims to develop basic number skills on the basis of informal strategies adults already possess and use outside school to handle everyday situations. This approach stands in rather stark contrast to the methods commonly employed in traditional mathematics classrooms, which have tended to heavily rely on the use of textbooks and workbooks and ignore the learner's informal knowledge.

In Chapter 8, Betty Hurley Lawrence introduces various forms of technology available for numeracy instruction, including "low tech" options such as audiotape and calculators as well as more advanced options such as personal computers and electronic communication networks. In discussing key applications of technology in mathematics instruction, Lawrence illustrates how these tools can provide adult learners with opportunities for self-directed and self-paced learning and more contextualized and realistic learning environments.

As Martha Sacks and Dorothy Cebulla recognize in Chapter 9, the many adult students with histories of learning difficulties present special challenges to ABE teachers, in mathematics as well as in other subject areas. The chapter explores a multifaceted approach to assessing the effects of potential perceptual, linguistic, and cognitive processing differences; it reviews instructional methods designed especially for adults whose prior experience suggests pervasive perceptual difficulties that affect their learning of mathematics; and discusses needed changes in teaching practices to enable learners with special learning needs achieve their full potential.

Closing this section is Chapter 10 in which Harriet Hartman synthesizes some of the theoretical and pragmatic perspectives presented by the other chapters in Section II, regarding issues of diversity deriving from gender, race and ethnicity, immigrant status, social class, and learning difficulties, thus offering readers a more integrated view of some of the issues dealt with separately in earlier chapters.

Section III, *Reflecting on Practice and Learning*, includes four chapters written by adult educators and researchers who reflect on innovative approaches to instruction they have developed. These chapters focus on the process of change teachers or their students have to manage, and raise penetrating questions and suggestions regarding the nature and value of topics and methods that are chosen for instruction.

The first two chapters in this section, Chapter 11 and Chapter 12, are written by Leslie Arriola and Donna Curly, respectively, who were both members of the Massachusetts ABE Math Standards Project, a group of teacher-researchers who examined and applied the NCTM curricular standards (discussed in chap. 3) to their adult learning contexts. These authors present their personal attempts to systematically reexamine their practice, evaluate the impact of introducing selected changes in their classrooms, and consider the implications for teacher research as a means of curricular and pedagogical reform in adult numeracy education.

In Chapter 13, Karen Hicks McCormick and Elizabeth Wadlington describe a process-writing approach they have developed to enable adult learners to devise, write, solve, and publish original "math stories" based on learners' own life experiences. The authors offer examples of student generated word problems, discuss the instructional strategies they employed, and describe the impact of this instructional approach on learners' conceptual understanding of mathematical topics as well as on their communication skills. The chapter serves to demonstrate an approach for integrating literacy and numeracy instruction that can respond to some of the challenges discussed by Gal in Chapter 1.

Based on her research on the mathematical practices embedded in everyday tasks, Mary Harris explores in Chapter 14 the richness of mathematics within needlework, the traditional work of women in many countries, through a series of case stories accompanied by mathematical analysis and suggested classroom activities. Harris suggests that, with proper adaptations, the idea of needlework as mathematics can help to reengage many women in mathematics learning and provide new and rich contexts for meaningful instruction. On a broader note, Harris examines the historical, social processes that have contributed to the gendering and de-evaluation of what she calls "Women's mathematic," thus opening a door into an important area not addressed elsewhere in this volume, that of the political and social forces that shape mathematics instruction and affect motivation and achievement.

Finally, *Section IV, Assessment,* draws on international perspectives to provide insights into progressive practices in assessing adults' numeracy skills.

In Chapter 15, Joy Cumming and Iddo Gal argue for the need to reform and broaden traditional approaches to assessment of numeracy skills in Adult Basic Education, which have tended to rely heavily on multiple-choice, computational, or standardized testing formats. They examine and demonstrate key shortcomings of traditional approaches, describe current "good practices" in the assessment of mathematical learning, and provide recommendations that can help educators and programs obtain necessary information about learners' performance and understanding.

The last chapter in this volume, by Dutch author Mieke van Groenestijn, provides a concrete and thought-provoking illustration of the potential for creating new and better forms of assessment, as advocated by Cumming and Gal and by other authors in this volume. van Groenestijn describes an innovative assessment method, the Supermarket Strategy, that was developed in The Netherlands and is integrated with the Realistic Mathematics curriculum used in both Adult Basic Education and K-12 education in that country. The author presents the principles of the Realistic Mathematics framework, outlines an adaptive testing strategy that uses familiar everyday materials as a basis for in-depth assessment of students' functional skills and reasoning strategies, and discusses implications for the needed changes in teachers' professional development.

I

PERSPECTIVES ON NUMERACY

1

The Numeracy Challenge

Iddo Gal
University of Haifa

This chapter provides an overview of the numeracy terrain and is organized in three parts. The first part explores the nature of numeracy and of numerate behavior in different situations. The second part elaborates on some aspects of numeracy, especially its cognitive and dispositional components, and on links between literacy and numeracy. The final part discusses implications and resulting challenges facing the adult numeracy community in four areas: emphasis on skill transfer and situation management capacities, integration of literacy and numeracy instruction, collaboration with K-12 systems, and support structures for numeracy education.

INTRODUCTION

Quantitative skills traditionally have been viewed as a basic skill area—one of the three R's that all people need to possess. Formal specifications of the kind of mathematical skills that adults may need, however, have often been vague. The United States National Literacy Act of 1991, for example, states that literacy is an "individual's ability to read, write and speak [in English], and to compute and solve problems at levels of proficiency necessary to function on the job and in society, to achieve one's goals, and develop one's knowledge and potential." Complementing this definition and providing more reasons for developing citizens' skills, one of the National Education Goals in the United States specifies, "Every adult American will be literate and will possess the knowledge and skills necessary to compete in a global economy and exercise the right and responsibilities of citizenship" (1998, p. VI).

Similar statements about desired skills and knowledge of citizens can be found in documents from other countries. Even when such statements recognize the need to attend to adults' quantitative skills, they are very broad, leaving educators, curriculum designers, and decision makers in all sectors of education to cope with the challenge of specifying what is implied by phrases such as "function . . . in society," "knowledge . . . necessary to compete in a global economy," "computational skills," or "problem solving." The challenge is complicated once we realize that adults need to manage *multiple* and *diverse* types of situations involving numbers, quantities, measurements, mathematical ideas, formulas, patterns, displays, probabilities and uncertainties, and events that unfold in time. Key examples are:

1. Home: Shopping, home repairs, cooking, coordinating schedules, understanding prescription labels;
2. Personal finance: Budgeting, filling tax forms, monitoring expenses, paying bills, negotiating a car loan, planning for retirement;
3. Leisure: Planning a trip or party, designing a crafts project, knitting;
4. Active parenting: Helping one's children with mathematics homework, understanding scores on standardized tests and statistics about the child's school;
5. Communicating with professionals: Talking with genetic counselors, obtaining medical advice, buying insurance;
6. Informed citizenship: Comprehending poll results discussed on TV or crime figures reported in a newspaper; writing a letter to a public official;
7. Social action: Helping with fund-raising or a survey for a local action group, debating environmental implications of a proposed development project;
8. Workplace: Shipping merchandise, measuring, computing materials needed, reading assembly instructions, retrieving data from a computer system, learning statistical process control, planning timetables;

9. Passing tests: Taking a college entrance exam or a technical certification test;
10. Further education: Studying college-level courses or taking technical training.

In these and other contexts, adults' mathematical know-how and numeracy in general serve multiple purposes. Together with literacy and other skills and knowledge, they promote access, orientation, and ability to keep up with a rapidly changing world. They enable or contribute to the expression of one's ideas and opinions and to effective participation in public life. They enhance independent functioning and action, coping with problems and dilemmas, and handling choices as a parent, citizen, or worker. They also serve as an important bridge (but also a gatekeeper) to further formal learning (Curry, Schmitt, & Waldron, 1996). However, no single agency or group has jurisdiction or control over the definition of the (mathematical or other) skills that adults may need for these diverse purposes or to be able to effectively manage the range of life contexts just illustrated.

Efforts to define aspects of the mathematical needs of adult life appear in publications representing perspectives of different players in the educational and policymaking arenas, such as:

Employers: A task force of the American Society of Training and Development (Carnevale, Gainer, & Meltzer, 1990), or later work by the Secretary of Labor's Commission on Achieving Necessary Skills (Packer, 1997; SCANS, 1991);

The K-12 education community: Willis (1990) in Australia, the National Council of Teachers of Mathematics (NCTM) (1989) and the American Association for the Advancement of Science (1993) in the United States, or the Cockroft report in the United Kingdom (1982);

The adult education community: Recent recommendations of a taskforce of the Adult Numeracy Practitioners Network (ANPN) in the United States (Curry et al., 1996);

Assessment initiatives: Large-scale surveys of adult literacy skills (Kirsch, Jungeblut, Jenkins, & Kolstad, 1993; Organisation for Economic Cooperation and Development [OECD], 1995; Wickert, 1989), programs for testing functional skills of adult learners (Rickard & Ackerman, 1994), or recent efforts in several European countries to standardize the diagnosis of skills of early school leavers, immigrants, or adults entering occupational training or adult basic education schemes (for a Danish example, see Wedege, 1996).

Many important commonalities can be found in these sources, such as the emphasis on the need for adults to have not only basic arithmetical knowledge, but good number sense, estimation skills, flexible problem-solving skills, or confidence in their knowledge, among others. Yet, in creating their recommendations, different agencies clearly employ different "lenses" and make somewhat different assumptions about the contexts in which people may need to function, and thus about needed knowledge and skills.

Although educators and planners may not presently agree as to the goals of formal education in mathematics for adults, many adults nonetheless do study mathematical topics as part of adult basic education (ABE), adult literacy classes, or workplace skills programs. In the United States, for example, a recent national survey (Gal & Schuh, 1994), based on a large-scale survey of 350 adult education programs in 15 states, has estimated that, out of almost 4 million adult students who were studying in publicly funded programs in the United States, over 80% received some math-related instruction.

Only in recent years have the adult education and adult literacy communities begun to pay visible attention to the mathematical side of adult education. The term *numeracy* is being increasingly used by members of these communities, yet its meaning is still elusive, and its implications for the desired content and process of educational efforts aimed at adults have not been broadly explored. Overall, this chapter aims to paint in broad strokes a picture of the numeracy terrain. The chapter is organized in three parts. The first part explores the nature of numeracy and numerate behavior in different situations. The second part elaborates on some aspects of numeracy and its links with literacy. The final part discusses implications and resulting challenges facing the adult education and adult numeracy communities.

NUMERACY AND NUMERACY SITUATIONS

Foerch (chap. 6, this volume) reminds us that we teach people, not mathematics. It follows that we have to first think about and create a vision of what we want people to learn—develop numeracy, become numerate—before we contemplate the means for achieving this goal (e.g., mathematics education). The term *numeracy* as used here describes an aggregate of skills, knowledge, beliefs, dispositions, habits of mind, communication capabilities, and problem-solving skills that individuals need in order to autonomously engage and effectively manage numeracy situations that involve numbers, quantitative or quantifiable information, or visual or textual information that is based on mathematical ideas or has embedded mathematical elements.

As with attempts to define literacy (Venezky, Wagner, & Ciliberti, 1990), what constitutes numeracy should be discussed in relative rather than absolute terms, in accordance with the goals and circumstances of a particular society or community. Like literacy, numeracy has multiple definitions, ranging from some that emphasize basic computational skills to those that encompass a broad and quite advanced range of skills and dispositions (Baker & Street, 1994).

How do schools go about preparing students to autonomously engage and effectively manage numeracy situations? Traditionally, school mathematics has emphasized an "internal" view of the goals of mathematics education; however, the internal view is presently in transition. The old view that mathematics instruction should aim to teach number facts and computational procedures is being gradually supplemented (or replaced) by a "new" view, one emphasizing problem solving, reasoning, communication, and the use of tools and group processes. This paradigm shift has been influenced by the revolution in cognitive science (Bruer, 1993) and by constructivist ideas in education; it is sometimes described by claiming that students should spend more time learning what mathematicians do (e.g., conjecture, experiment, check hypotheses, verify results, explain) rather than what mathematicians know (e.g., number facts, computational rules, formulas, proofs).

Although this new internal view offers exciting possibilities for improving learning and achievement in mathematics, it still pays insufficient attention to an "external" view of what people (rather than mathematicians) need to be able to do, and how people actually manage their affairs beyond the school walls. The goal of developing "mathematical power," introduced by NCTM (1989; see chap. 3), is an important step in this direction. The next sections elaborate on the nature of numeracy situations and the notion of the management of such situations.

Situations

Real-life numeracy situations are always embedded in a life stream with real, personal meaning to the individual involved. The following are three key examples (with some subtypes) that illustrate the range of numeracy situations.

Generative situations require actors to count, quantify, compute, or otherwise manipulate numbers, quantities, items, or visual elements, and eventually create (generate) new numbers. Such tasks involve language skills to varying degrees. Two important and interrelated subtypes of generative situations are computational tasks and quantitative literacy tasks.

Computational tasks normally demand the generation of a single number. Individuals apply the four basic operations or other simple counting or arithmetical procedures to numbers or quantities clearly evident in the situation. The figure computed can be clearly marked as "right" or "wrong," regardless of the process by which it was reached. An example is calculating the total price of products when shopping, finding the number of boxes in a crate, or finding the area of a room that has to be carpeted or painted in order to calculate the amount of materials needed. (Note that computational tasks are the most frequent type of tasks found in math textbooks for adults).

Quantitative literacy tasks require that people apply arithmetic operations to information *embedded in written materials*, such as when necessary data has to be extracted from different forms or documents, computational operations have to be inferred from printed directions, or quantitative arguments to be comprehended are embedded in technical documents or newspaper prose. Such tasks were described by Kirsch and his colleagues (e.g., Kirsch et al., 1993) as one of three facets of literacy (the others being prose literacy and document literacy), in the context of the measurement of literacy levels in the general population in large scale surveys. Examples are reading a menu and computing the cost of a specified meal, filling out an order form for a product, reading a bus schedule to figure out travel times between stations, or comparing which of two medical benefits packages is better based on information in brochures.

Interpretive situations demand that people make sense of, and grasp the implications of, verbal or text-based messages that may be based on quantitative data but that *do not involve direct manipulation of numbers*. Such situations may involve no numbers at all but still refer to important ideas that are part of mathematics or statistics, as when samples, bias, correlation, or causality are implied. The response expected of an actor in such a situation is often the creation of an opinion or the activation of a set of critical questions to be answered before the information or arguments presented are accepted as credible, sensible, or valid. Such opinions or answers to critical questions cannot necessarily be classified as "right" or "wrong" (as with responses to generative tasks), but rather are judged in terms of factors such as their reasonableness or the quality of the arguments or evidence on which they are based (Gal, 1998). An example is being faced, when reading a newspaper or watching news on TV, with a report of results from a recent local poll or from a small-sample medical experiment, possibly involving references to percentages, averages, rate changes, or generalizations about differences between population groups.

Decision situations demand that people find and consider multiple pieces of information in order to determine a course of action, typically in the presence of conflicting goals, constraints, or uncertainty. Two key subtypes here are optimization tasks, which require the identification of optimal ways to use resources such as money or supplies, or schedule personnel or time (see SCANS, 1991); and, choice tasks, which require a choice among alternatives (see chap. 4), as when people have to decide which of several apartments they should rent. As with interpretive situations, a response to a decision situation will have to be evaluated in terms of its reasonableness; yet, a larger subjective factor is involved, as a response to a decision situation will be formed in part based on elements such as the assumptions made about future trends, judgments of event probabilities, or the actor's preferences and value system. An example is when a family has to compare information from several banks to decide which loan schedule is the "best" and most manageable, or whether to purchase a certain type of insurance.

The three types of numeracy situations described above should not be viewed as distinct task categories but as clusters in a hypothetical "numeracy task space" defined by dimensions such as the nature of the required response (e.g., generative as opposed to interpretive), the number and characteristics of the quantitative elements in the situation, or the extent and nature of the literacy processes involved. Clearly, other types and hybrid cases of such numeracy situations are possible.

The nature of the skills and dispositions that an actor may need for managing a situation will depend on the demand characteristics of the situation and on the actor's goals just as much as on the type of data or quantitative information available in a situation. For example, given the same material, such as a bus schedule, a situation may be interpretive for one actor yet decisional for another, or change its nature as time progresses. Likewise, given information in a newspaper article (e.g., text with accompanying graph), an actor may attempt, as Curcio (1987) describes, to literally read the data (i.e., find numbers in the graph), read "between the data" (i.e., compare data points, find differences), or read "beyond the data" (i.e., conjecture, extrapolate), depending on the circumstances.

Generative situations and their subtypes have traditionally been the main focus of math instruction for adults (and for school students, for that matter), and in textbooks have been simulated primarily by simple word problems. The use of simple generative tasks may be sensible if done as a part of formal school-type mathematics learning or exam preparation. Is it sufficient, however, when the educational goals involve the preparation of students for the full range of numeracy situations in the real world? Complex or literacy-rich generative situations, and especially decision and interpretive situations, have been and still are poorly represented in teaching resources or addressed by instructional efforts.

With this in mind, and given that math instruction has traditionally focused primarily on selected generative tasks, we should ask, How prepared are adults (or school graduates) to effectively handle the many diverse generative tasks that may come up in the real world? As for adults in the United States, roughly 50% of the thousands of individuals interviewed for the recent National Adult Literacy Survey (Kirsch et al., 1993), including many adults with high school and postsecondary credentials, had major difficulty with diverse "quantitative literacy" tasks such as those described earlier. Results of a similar nature were obtained in surveys in other industrialized countries (OECD, 1995). There may be several explanations for such findings (Dossey, 1997), yet issues related to the nature and acquisition of functional as opposed to formal knowledge are surely involved.

Managing Numeracy Situations

Most school mathematics problems normally have knowable "right" answers and their solutions can be verified by the teacher. Yet, many numeracy situations are not resolved with "solutions" that can be classified as right or wrong. In fact, adults do not "solve" situations (in contrast to how students solve word problems, which are supposed to simulate real-life situations). Rather, adults manage situations. When faced with a real-life numeracy situation, adults identify and choose one of several courses of action, based on considerations of personal goals, situational demands, severity of consequences, personal and situational resources, and so forth. Faced with a generative task, an actor may ask, for example:

- How accurate should I be?
- Can I afford to make mistakes?
- What resources and how much time do I have?
- How much energy can I spend on this?
- What is the cost of being inaccurate or slow?
- What aids (e.g., calculators, people) can I use?
- Should I get any help? Why or why not?
- Can I get somebody to do it for me, and how much can I trust them to do it right?

Based on such considerations, adults will engage in managerial activities, such as planning a response to a situation, executing plans and monitoring progress, revising a course of action, or making new plans. This means that even in a situation that a traditional math teacher would classify as computational, an actor may find it fitting to sacrifice the precision or quality of his or her response (e.g., by estimating a tip rather than calculating exactly), aim for a safety margin (e.g., by overestimating the amount of food to order for a party), or delegate responsibility as needed (e.g., ask a family member or a salesperson for help). The response may be reached in a computationally inefficient way or be based on nonstandard or invented procedures, but this may not matter as long as the actor manages the situation in a way that is reasonable in light of the demand characteristics of the situation, social conventions, and the actor's short- and long-term goals. Further, like managers, adults have to adapt to changing circumstances and be able to identify trends that may develop into problems in the *future* (e.g., cash flow, medical condition); these require that adults can initiate reasoning and anticipatory cognitive processes. (For an interesting example of such adaptive processes in the area of industrial mathematics, see Pollak, 1997).

ELABORATING ON ASPECTS OF NUMERACY AND ITS DEVELOPMENT

The use of numeracy (as a parallel term to literacy) intends to help us keep in mind that the range of skills and dispositions required for effective functioning in most life contexts is much wider, and often quite different, than what has been traditionally addressed in K-12 mathematics education. It also reminds us that in many life situations adults need to seamlessly integrate the use of mathematical and literacy skills. To help learners become able to autonomously engage and effectively manage numeracy situations, we need to attend to both the cognitive and dispositional components of numeracy. The cognitive component refers to the skills and knowledge that learners should develop and become able to apply in diverse situations. Equally important, however, is the dispositional component—the beliefs, attitudes, and "habits of mind"—that learners need to develop in relation to their growing knowledge and skills and that are needed to support effective management of numeracy situations.

Cognitive/Skill Aspects of Numeracy

What specific skills and knowledge should be emphasized in numeracy education? Most educators would likely agree that adults living in industrialized societies should possess some knowledge in key areas of mathematics, such as basic arithmetical knowledge and facility with number operations, measurement, basic geometry, basic algebra, and interpretation of basic graphs. Other areas receiving growing recognition include, for example, number sense, estimation and "mental math" skills, flexible problem-solving skills, and the ability to make judicious use of calculators. (Other chapters in this volume discuss these and other essential skill and knowledge areas in more detail.)

The NCTM *Curriculum Standards* (NCTM, 1989; and the upcoming Standards 2000 publication) expand the scope and nature of what mathematics teaching and learning may entail (at least in K-12 contexts) well beyond traditional notions of drill-and-practice instruction. The *Standards* position the development of "mathematical power" as one of the key overarching goals of mathematics education and encourage a view of mathematics learning that requires and aims to develop reasoning, problem solving, communication, and connections between knowledge and skill areas.

Recently, a taskforce of the Adult Numeracy Practitioners Network (ANPN) in the United States presented a framework describing numeracy skills needed to "equip learners for the future" (Curry et al., 1996). Building on the NCTM *Standards*, this project identified seven themes that should serve as foundations or standards for adult numeracy education, as well as a corollary need for adults to become confident in their knowledge:

- Relevance/Connections
- Problem Solving/Reasoning/Decision Making
- Communication
- Number and Number Sense
- Data, Statistics, and Probability
- Geometry: Spatial Sense and Measurement
- Algebra: Patterns and Functions

A "numeracy learning space" or course matrix can be defined by crossing these seven content themes (perhaps with a specification of subareas) with the purposes for which such knowledge is needed. One illustrative example for this approach is a framework presented by a team of Spanish adult educators from the Escuela Popular de Oporto in Madrid (Plaza, 1996). These educators define eight purposes for knowing mathematics: everyday life; interpretation of information; the world of work; being a consumer; health; dealing with technologies; environment; and social justice, solidarity, and life in a democracy. In addition, they specify eight basic mathematical content areas: knowledge of numbers; four basic operations; estimation and mental arithmetic; measurement and handling of units; proportions; diagrams, statistics and probability; and handling of calculators. Specific real-world tasks are described by Plaza (1996) for each of the resulting 64 cells in the learning matrix, making this a useful roadmap for educators. (The reader may want to compare the eight Spanish purposes with the list of 10 numeracy contexts presented in the Introduction of this chapter.)

Specification of content areas or learning matrices such as those just described can enhance our understanding of the skills that adults may need within different cultural, social, or economic contexts. Yet, for several interrelated reasons it may be difficult or impossible to reach full consensus about the types and levels of mathematical skills to be emphasized in any one numeracy program.

First, conceptions of numeracy may be specified from different perspectives, for example, in terms of knowledge of specific content areas in mathematics (see earlier), or as functional skills for handling various everyday or workplace tasks. The skills required for access to higher or further education, in which formal knowledge of mathematics and mastery of relatively advanced topics such as algebra or calculus are often required, are not necessarily those desired by employers or those that adults see as useful in their own everyday life. Adult educators are often frustrated when they have to prepare students for taking "gatekeeper" tests because the focused content and formal skills required by such tests may conflict with broader conceptions they may have of what adult numeracy may encompass, and both of these may clash with learners' own expectations.

Second, choices have to be made between the many options for extending a basic curriculum, each with its own rationale (e.g., functions, calculus, variability in natural or manufacturing processes, modeling, scientific reasoning, design of experiments and surveys, financial literacy, using electronic spreadsheets). It is unfeasible to assume all topics can be covered in any given program of learning.

Third, as suggested earlier, numeracy should be discussed in relative terms: being numerate, even in a particular community, may involve different subsets of skills for different people. The routes to becoming numerate may take different trajectories, depending on personal goals or societal circumstances. Thus, it may be unrealistic to define a single ladder with fixed steps on which all learners are expected to climb, as has been the case with goals specified for mathematics teaching in many K-12 school systems.

Last, some specific content areas may be acquired hierarchically and must be taught sequentially, but others can be developed relatively independently of each other or taught with no particular sequence in mind, allowing for different learning paths. With that in mind, and because many adult students participate in learning for only a limited length of time, adult educators may have to chart individualized routes for visiting only selected "cells" in a matrix, taking into account the expected contexts in which the acquired knowledge and skills will be used, and the need to develop generalized problem-solving, reasoning, and communication capacities that can support transferrable skills.

Dispositional Aspects of Numeracy

How well a numeracy situation is managed depends not only on the technical know-how of its manager (e.g., knowledge of mathematical rules and operations, linguistic skills), but on his or her beliefs, attitudes, metacognitive habits and skills, self-concept, and feelings about the situation. People's dispositions may affect the way they approach or react to learning episodes in a numeracy classroom, or how they respond to and manage numeracy situations outside the classroom.

Many people who come to adult education programs report negative attitudes about learning math or addressing everyday mathematical tasks. Some adults, including highly educated ones, decide that they are not "good with numbers." These sentiments or perceptions are usually attributed to the negative prior experiences they have had as students of mathematics (Tobias, 1993), and stand in contrast to the desired sense of "at-homeness with numbers" (Cockroft, 1982). Such attitudes and beliefs can and often do interfere with the students' motivation to develop new mathematical skills or to tackle math-related tasks and may also affect test performance. As a result, instruction often needs to overcome learners' prior "math abuse" (Johnston, 1992) and "undo" cognitive habits, unproductive beliefs about the relevance of mathematics to real-life, expectations about what learning and applying mathematical knowledge should look like, and negative self-perceptions that learners (more often women than men) bring to the classroom (McLeod, 1992).

In realistic contexts, adults with a negative mathematical self-concept may elect to avoid a problem with quantitative elements, address only a portion of it, or prefer to delegate or subcontract a problem, for example, by asking a family member or a salesperson for help. Such decisions or actions are indeed the prerogative of a manager and can serve to reduce both mental and emotional load. Yet, such actions may fall short of autonomous engagement, carrying negative consequences, especially not being able to fully achieve one's goals.

Adults with negative dispositions may have long-standing habits such as depending heavily on assistance from others, or sticking to informal (and possibly less efficient and accurate) methods. Showing students that "math is fun" (e.g., by using math games) may be of some help in reversing negative feelings about school math, but may not be sufficient to cause students to change entrenched beliefs (Schoenfeld, 1992) or habitual reactions to real-life situations that involve mathematical elements.

A different facet of people's dispositions is related to their metacognitive habits. In interpretive situations, for example, we want adults to be aware of critical questions that should be raised (e.g., about the credibility of the source of a message, about sample size or adequacy of sampling procedures used in a survey). We also want them to foster a *critical stance*, which involves a propensity to spontaneously invoke, *without* external cues, the list of critical questions, and further invest the mental effort needed to ask penetrating questions and try to answer them. Without this stance, people might accept objectionable arguments and develop an incorrect world view.

It follows that a basic goal of numeracy education is to bring students to a point at which they are able to positively approach situations with quantitative elements and to help them appreciate the costs of engaging in actions without taking full account of all the [quantitative] information given and its various; that is, if they operate impulsively, are uncritical, or do not employ what Baron (1988) calls "active open-minded thinking." The upshot is that conceptions of numeracy should address not only issues that are purely cognitive, but also students' dispositions and cognitive style.

Links Between Literacy and Numeracy

The acquisition, teaching, and learning of language skills and of mathematical skills have been mostly addressed by teachers as well as the professional literature as two separate areas of inquiry and practice with little crossover. However, there are several related intersections between literacy and numeracy.

Mathematics as a language. Mathematics can be viewed as a separate language system, with its own symbols, vocabulary, syntax, grammar, and semantics (Halliday, 1979). The language of mathematics can be used to describe or model situations and to communicate both concrete and abstract descriptions and ideas. Mathematics appears as a language when we examine, for example, the process of learning or using a formula, which involves *reading* each element of a formula, *comprehending* the meaning of each element or term, and *constructing* a sense for the intention of the whole formula. Further, learners have to be able to make sense of whole phrases (or visual displays) stated in the language of mathematics, extract their meaning, and manage cases when alternate syntactical or lexical forms are used (e.g., "What do 9 plus 12 sum to?"

"Find the total of 9 and 12"). In addition, the expression of mathematical ideas depends in part on a person's natural language, a situation that can create difficulties for adults who are fluent in one language yet try to learn how to "speak mathematically" in another language.

Language factors in learning mathematics. Language, in both oral and written forms, is the prime medium through which the learning of mathematics is mediated in either formal or informal schooling. Students have to read and decode written mathematical terms or elements, as well as comprehend the action implications of these elements (Laborde, 1990). They have to adapt to communicative conventions and local vocabularies set forth by their teacher or textbook and become aware that the meanings of terms used in a math classroom are often more constrained, precise, or plainly different than when terms (e.g., table, average, minus, group, proof, volume) are used in everyday speech. Also, students may find mathematics textbooks difficult to read and word problems in them difficult to parse and solve. As Kane, Byrne, and Hater (1974) argue, the terse nature of mathematical texts requires strategies for reading and monitoring comprehension that only partially overlap those used in dealing with ordinary text (e.g., prose). Finally, students are increasingly expected to be able to effectively communicate with peers and teachers through verbal and written means (e.g., Sterrett, 1990). Teachers are encouraged to ask students to talk and write about problem-solving processes, describe procedures, clarify conjectures, voice observations about given data, or explain results or reasoning processes, in part by means such as writing journals, preparing project reports, or writing their own math problems (see chaps. 10 and 12). This emphasis on communication is used both to support the learning process, as well as because of the realization that communicative acts are part of the fabric of many real-world numeracy situations.

Language-mathematics links in real-world contexts. As pointed out earlier, one way to characterize numeracy situations is by the extent to which they are generative as opposed to interpretive. Some generative tasks may involve few if any language skills, whereas others, especially quantitative literacy ones, will require stronger literacy skills and may blend elements of interpretive skills. Effectively, management of "pure" interpretive tasks involves reading, writing, language comprehension, and other literacy skills to a much greater degree; it also requires solid familiarity with the context of the task, conceptual understanding, and a critical stance, rather than only computational prowess.

DISCUSSION AND IMPLICATIONS

The term *numeracy* as used here captures a terrain that is broader, more diverse, and more functional in nature than that which has been traditionally targeted by mathematics education, at least for adults. The range and characteristics of numeracy situations, and their complex and dynamic nature, suggest that there are important differences between "knowing mathematics" in a classroom context and being able to demonstrate numerate behavior in different situations. Numerate behavior is shaped by personal needs and goals as well as by social processes and emerges out of the interaction between cognitive and dispositional processes and situational circumstances.

In addition, a discussion of what it means to be numerate—in the classroom or in the real world—must consider a person's literacy and communication skills. There are significant areas where literacy and numeracy blend into a single skill. However, there are also many quantitatively rich, real-world situations (and definitely many classroom situations) that require constrained or minimal use of language skills. Therefore, numeracy may have to be viewed as semi-autonomous from literacy. Figure 1.1 depicts this hypothesized relationship between literacy, mathematics, numeracy, and general problem-solving and situation-management skills.

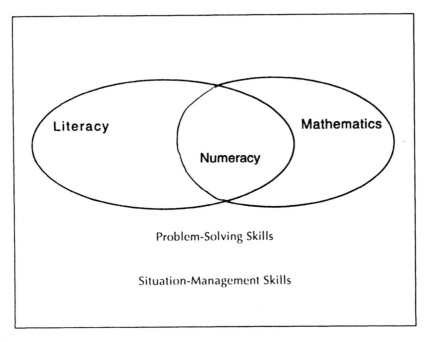

Figure 1.1. Numeracy in relation to other skill domains

The expectation that adults will be able to *manage* effectively *multiple* types of numeracy situations, including many that require an integrated application of literacy and numeracy skills, presents many challenges to the adult education community and to decision makers at local and national levels. Four sets of challenges are discussed next.

Attend to skill transfer and "situation management." A first set of challenges stems from the need (a) to understand the factors and processes affecting learners' ability to transfer skills, whether these were acquired in or out of the classroom; and (b) to enhance learners' ability to effectively manage diverse numeracy situations that are different from those encountered during instruction. The range of numeracy situations in work, daily life, and civic contexts—and hence the range of skills and dispositions required for their effective management—is wider and different in important ways from that which has traditionally been addressed in mathematics classrooms. For instance, in her ethnographic study of carpet-layers, Masingila, Davidenko, and Prus-Wisniowska (1996) found that the way these workers handle many of their tasks is quite different from how these tasks are portrayed in textbook word problems about carpet laying. Therefore, the extent to which instructional methods and assessment tasks simulate complex, real-life events that require true "situation management" (in contrast to solving word problems) should be carefully reviewed. A similar conclusion can be drawn from a recent study by Pozzi, Noss, and Hoyles (1998), who studied mathematical activities of pediatric hospital nurses.

Some educators may believe that traditional modes of math instruction successfully enable students to develop solid basic skills, habits of mind, and logic that are sufficient to manage a wide range of numeracy situations outside the classroom. In this regard, some teachers may argue that the goal of instruction should be to help students discover and understand the abstract principles underlying topics such as geometry or algebra, for example, and that the process of learning to understand such principles makes students into good thinkers who can in turn apply their "thinking tools" in almost any situation. Although this may indeed hold for some students (usually the more advanced ones), a growing literature suggests that people and students of all ages often have trouble transferring and applying mathematical skills in new contexts, in or out of the classroom (Ginsburg & Gal, 1995; Schliemann, 1998). Such findings corroborate informal observations familiar to many teachers: too many students have trouble dealing with presumably standard problems that they just "successfully" solved, once minor aspects of the problems are changed.

There is a growing recognition of the educational benefits of embedding learning in meaningful contexts and of connecting knowledge with its applications in authentic settings (e.g., Mikulecky, Albers, & Peers, 1994; SCANS, 1991; Wolf, 1990). Psychological and educational research suggests that cognition is often "situated," that is, knowledge and problem solving are contextualized and are not constructed entirely from static representations in memory that can be activated and applied to any situation (Brown, Collins, & Duguid, 1989; Derry & Lesgold, 1996; Singley & Anderson, 1989). It follow that instruction should blend functional/problem-oriented and "pure" mathematics topics and incorporate designs to maximize the applicability and transferability of knowledge gained in the classroom to everyday and work environments. Options that may enhance such skill transfer are, for example, the use of multimedia-based simulations in-classroom role playing (see chap. 8) extended classroom projects, or in-basket methods.

Acceptance of the notions advocated here of numeracy and the centrality of situation-management skills should impact the types of assessment tasks that teachers use in the classroom or that employers or testmakers present to candidates. Assessments employed for either formative or summative purposes should be designed to provide, among other things, useful information about learners' ability to transfer their skills and manage numeracy situations of interest to the student and the program (see chaps. 15 and 16). Yet, assessment users should keep in mind possible limits on the interpretations that can be attached to a person's performance in simulated situations of different kinds. Overall, the use of functional, authentic tasks in instruction and assessment can benefit all students, whether their primary goal is to learn to cope with formal mathematics, as is often the case in some adult education contexts, or to manage other numeracy situations.

Further, we are challenged to ask what foundations should adult numeracy lay, so that students are *willing* to invest in either *learning* new (mathematical) skills or *using* their skills and knowledge, as required by changing life circumstances? Dispositional issues, although difficult to fit into conceptions of numeracy development that are couched in cognitive terms, are of critical importance (McLeod, 1992). Numeracy education should present quantitative reasoning as a viable way to approach life's challenges, in order to increase the likelihood that learners feel confident to engage numeracy situations (see Gal, Ginsburg, & Schau, 1997, for an illustrative examples related to statistical literacy). Numeracy education should serve as a gateopener, instead of a gatekeeper; learners, after leaving a program, should be motivated to further develop their numeracy skills and engage in lifelong learning through either formal or informal means (Benn, 1997).

The NCTM *Standards* and the reform movement that followed it emphasize that the view of "mathematics as problem solving" should be a key process standard in mathematics education. This emphasis has been pivotal in pointing the attention of teachers and curriculum developers to the importance of problem-solving issues. Yet, it may inadvertently perpetuate the belief that development of numerate people can occur entirely in the (mathematics) classroom, independent of the larger present *and* future context of learners' lives.

Given an "external" view, the NCTM standard of "mathematics as problem solving" should be reformulated or supplemented with "solving problems with mathematics" (with the emphasis that problems and numeracy situations, not just mathematics, should be the driving force for learning and teaching). More broadly, teaching focused on problem solving should develop skills and dispositions needed for the "autonomous management of numeracy situations." These suggestions follow from three realizations: (a) realistic problem situations may or may not have to be approached only by the mathematical tools and limited range of reasoning and problem-solving strategies practiced in school; (b) the criteria for judging the quality of the "solution" of any problem do not have to be couched only in the mathematical domain, as is clearly the case with many decisional and interpretive tasks; and (c) mathematics and literacy are intertwined, hence problem solving should not be viewed solely as a mathematical activity.

Integrate literacy and numeracy instruction. As discussed earlier, many numeracy tasks, especially interpretive tasks and other tasks involving numbers or quantitative statements embedded in text (e.g., forms, schedules, manuals, technical and financial documents), require adults to seamlessly integrate the use of numeracy and literacy skills. The instructional implications of this fact have hardly been explored, even though this skill integration is critical in enabling adults to be "smart" consumers and informed citizens. Classroom instruction too often perpetuates the separation between mathematical and other aspects of literacy education. Textbooks and other instructional resources use distilled, parsimonious language that often does not replicate the diverse types of real-world texts and does not represent the communicative demands found outside the school.

Teachers, administrators, and curriculum developers need to acknowledge that literacy and numeracy are inextricably connected and explore ways in which the development of people's literacy skills can also be promoted through instructional experiences seemingly more related to numeracy, and vice versa. In contrast, the current situation is that both areas of education compete for limited classroom time and teacher attention. Given that literacy instructors often end up having to teach mathematical issues as well (Gal & Schuh, 1994), it is essential to review the ways in which literacy and numeracy instruction is handled within the practice of a single teacher, or integrated across subjects or classes in a program.

The introduction of the "communication standard" by NCTM (1989) placed much needed emphasis on the importance of attending to both oral and written communication issues. Yet, teachers may still be more willing to adopt communicative tasks (e.g., student journals, project logs) that support an internal view of the goals of teaching mathematics than tasks that are based in real-world numeracy situations. Classroom and assessment activities will have to be further expanded to represent the full range, complexity, and diversity of the kinds of texts and literacy demands that adults have to manage in the world.

Strengthen collaborations with K-12 systems. A third set of challenges relates to the need to improve the connections between the adult numeracy system and K-12 mathematics education, for the benefit of both enterprises. The sentiment that "child" stuff or resources for teaching school mathematics are of little value for adult numeracy educators can and should change once it is realized that all communities involved in math education share some common goals. Adult educators should reap the outcomes of efforts to reform mathematics education at the primary and secondary levels, currently underway in many countries. Clearly, many resources and programs originally developed in K-12 contexts, such as those related to manipulatives, calculators, cooperative learning, computer-based simulations, alternative assessments, writing in mathematics classrooms, or family math, among many others, can be adapted for use with adult students. Likewise, professional development programs for K-12 teachers may be instrumental for the many adult educators who engage in teaching mathematics yet lack a solid base in relevant teaching methods or a deep understanding of mathematical subjects.

From a different angle, the adult education community is the "receiving dock" for many people who "fell through the cracks" of the K-12 system or were subjected to various forms of math abuse (Tobias, 1993). Mechanisms should be sought for sharing with K-12 educators the lessons learned by adult educators about unproductive modes of thinking or dispositions observed in their students, so as to help in preventing such phenomena and reducing their future incidence.

Finally, few if any conceptual and instructional links seem to have been established between family literacy initiatives in adult education and family math initiatives in K-12 education (such as the Lawrence Hall of Science Family Math program; see Stenmark, Thompson, & Cossey, 1986). Closer ties and collaboration are warranted, as both communities are involved in preliminary attempts to impact the intergenerational transfer of practices and beliefs.

Improve support structures. A final set of challenges relates to the need to improve the support and infrastructure for numeracy provision. Stronger academic recognition is essential to facilitate research focused on adult learners in general and particularly on development of numeracy-related skills and dispositions in both formal and informal settings. Also, further professionalization of adult numeracy practitioners is essential, as has been pointed by several working groups (e.g., Gal, Schmitt, Stoudt, & Ginsburg, 1994). In some countries, such as Australia, Holland, or England, standard workshops and collections of teaching materials have been developed for practitioners. Intensive attention in Australia to numeracy issues, for example, has led to the creation of an array of teaching resources as well as teacher training packages and modules that can be used locally by programs (e.g., Tout & Johnston, 1995).

Stronger international ties between numeracy practitioners and researchers are needed to facilitate solutions to common issues, such as the teaching of numeracy skills to new immigrants, to students with learning differences, or as part of family-oriented programs. Important beginnings can be found in professional forums (see chap. 2) as well as in recently established Internet sites and e-mail discussion groups focused on numeracy and math education in general. Yet, a challenge facing the adult numeracy community in virtually all countries is getting academic institutions to offer graduate courses related to adult numeracy teaching. Such courses, which presently are almost nonexistent, can combine essentials of the knowledge gained in K-12 math education contexts with those aspects unique to adult learning in general and adult numeracy in particular, as described in this book.

Overall, the combination of better and broader research, academic recognition, and strengthened professional development opportunities, is essential to the formation of informed and reflective practitioners who can adapt their practices to match the unique goals, needs, backgrounds, skills, and dispositions of all numeracy learners.

REFERENCES

American Association for the Advancement of Science (Project 2061). (1993). *Benchmarks for science literacy.* New York: Oxford University Press.

Baker, D., & Street, B. (1994). Literacy and numeracy: Concepts and definitions. In T. Husen & E. A. Postlethwaite (Eds.), *Encyclopedia of education* (Vol. 6, pp. 3453-3459). Oxford, England: Pergamon Press.

Baron, J. (1988). *Thinking and deciding.* New York: Cambridge University Press.

Benn, R. (1997). *Adults count too: Mathematics for empowerment.* Leicester, UK: NIACE.

Brown, J. S., Collins, A., & Duguid, P. (1989). Situated cognition and the culture of learning. *Educational Researcher, 18*(1), 32-42.

Bruer, J. (1993). *Schools for thought: A science of learning in the classroom.* Cambridge, MA: MIT Press.

Carnevale, A. P., Gainer, L. J., & Meltzer, A. S. (1990). *Workplace basics: The essential skills employers want.* San Francisco: Jossey-Bass Publishers.

Cockcroft, W. H. (1982). *Mathematics counts.* London: Her Majesty's Stationary Office.

Curcio, F. R. (1987). *Developing graph comprehension: Elementary and middle school activities.* Reston, VA: National Council of Teachers of Mathematics.

Curry, D., Schmitt, M. J., & Waldron, W. (1996). *A framework for adult numeracy standards: The mathematical skills and abilities adults need to be equipped for the future.* Final report from the System Reform Planning Project of the Adult Numeracy Practitioners Network. Washington, DC: National Institute for Literacy.

Derry, S., & Lesgold, A. (1996). Toward a situated social practice model for instructional design. In D. C. Berliner & R. C. Calfee (Eds.), *Handbook of educational psychology* (pp. 787-806). New York: Macmillan.

Dossey, J. A. (1997). National indicators of quantitative literacy. In L. A. Steen (Ed.), *Why numbers count: Quantitative literacy for tomorrow's America* (pp. 45-59). New York: The College Board.

Gal, I. (1997). Assessing statistical knowledge as it relates to students' interpretation of data. In S. Lajoie (Ed.), *Reflections on statistics: Agendas for learning, teaching, and assessment in school contexts* (pp. 275-295). Mahwah, NJ: Erlbaum.

Gal, I., Schmitt, M. J., Stoudt, A., & Ginsburg, L. (1994). Summary of conference recommendations. In I. Gal & M. J. Schmitt (Eds.), *Proceedings of the 1994 National Conference on Adult Mathematical literacy* (pp. 1-18). Philadelphia: University of Pennsylvania.

Gal, I., & Schuh, A. (1994). *Who counts in adult literacy programs? A national survey of numeracy education* (Tech. Rep. No. TR94–09). Philadelphia: University of Pennsylvania, National Center on Adult Literacy.

Gal, I., Ginsburg, L., & Schau, C. (1997). Monitoring attitudes and beliefs in statistics education. In I. Gal & J. B. Garfield (Eds.), *The assessment challenge in statistics education* (pp. 37-54). Amsterdam: IOS Press.

Ginsburg, L., & Gal, I. (1995). Linking informal knowledge and formal skills: The case of percents. In D. T. Owens, M. K. Reed, & G. M.

Millsaps (Eds.), *Proceedings of the 17th Annual Meeting of the North American Chapter of the International Group for Psychology of Mathematics Education* (pp. 401-406). Columbus, OH: ERIC Clearinghouse for Science, Mathematics, and Environmental Education.

Halliday, M. A. K. (1979). *Language as social semiotic: The social interpretation of language and meaning.* London: Edward Arnold Publishers.

Johnston, B. (Ed.). (1992). *Reclaiming mathematics.* Canberra, Australia: Department of Employment, Education and Training.

Kane, R. B., Byrne, M. A., & Hater, M. A. (1974). *Helping children read mathematics.* New York: American Book Company.

Kirsch, I. S., Jungeblut, A., Jenkins, L., & Kolstad, A. (1993). *Adult literacy in America: A first look at the results of the National Adult Literacy Survey.* Washington, DC: National Center for Education Statistics, U.S. Department of Education.

Laborde, C. (1990). Language and mathematics. In P. Nesher & J. Kilpatrick (Eds.), *Mathematics and cognition* (pp. 53-69). New York: Cambridge University Press.

McLeod, D. B. (1992). Research on affect in mathematics education: A reconceptualization. In D. A. Grouws (Ed.), *Handbook of research on mathematics teaching and learning* (pp. 575-596). New York: Macmillan.

Masingila, J. P., Davidenko, S., & Prus-Wisniowska, E. (1996). Mathematics learning and practice in and out of school: A framework for connecting these experiences. *Educational Studies in Mathematics, 31*(1-2), 175-200.

Mikulecky, L., Albers, P., & Peers, M. (1994). *Literacy transfer: A review of the literature* (Tech. Rep. TR94-05). Philadelphia: University of Pennsylvania, National Center on Adult Literacy.

National Council of Teachers of Mathematics (NCTM). (1989). *Curriculum and evaluation standards for school mathematics.* Reston, VA: Author.

National Education Goals Panel (1998). *The national education goals report: Building a nation of learners, 1998.* Washington, DC: U.S. Government Printing Office.

Organisation for Economic Cooperation and Development (OECD). (1995). *Literacy, economy, and society.* Ottawa, Canada: Statistics Canada.

Packer, A. (1997). Mathematical competencies that employers expect. In L. A. Steen (Ed.), *Why numbers count: Quantitative literacy for tomorrow's America* (pp. 137-154). New York: The College Board.

Plaza, P. (1996). *Materials from the adult's school in Madrid.* Poster presented at Working Group 18 (Adults returning to mathematics education), Eighth International Congress on Mathematics Education, Seville, Spain.

Pollak, H. O. (1997). Solving problems in the real world. In L. A. Steen (Ed.), *Why numbers count: Quantitative literacy for tomorrow's America* (pp. 91-105). New York: The College Board.

Pozzi, S., Noss, R., & Hoyles, C. (1998). Tools in practice: Mathematics in use. *Educational Studies in Mathematics, 36*(2), 105-122.

Rickard, P., & Ackerman, R. (1994). Comparison of literacy to numeracy achievement in a life-skills context. In I. Gal & M. J. Schmitt (Eds.), *Proceedings of the 1994 National Conference on Adult Mathematical Literacy* (pp. 33-40). Philadelphia: University of Pennsylvania.

Schliemann, A. (1998, July). *Everyday mathematics and adult mathematics education* (Keynote address). Proceedings of the Fifth International Conference of Adults Learning Maths: A Research Forum, University of Utrecht, The Netherlands. (A publication of Goldsmith Collete, University of London. London: Avanti Books).

Schoenfeld, A. H. (1992). Learning to think mathematically: Problem solving, metacognition, and sense making in mathematics. In D. A. Grouws (Ed.), *Handbook of research on mathematics teaching and learning* (pp. 334-370). New York: Macmillan.

Secretary of Labor's Commission on Achieving Necessary Skills (SCANS). (1991). *What work requires of schools: A SCANS report for America 2000.* Washington, DC: U.S. Government Printing Office.

Singley, M. K., & Anderson, J. R. (1989). *The transfer of cognitive skill.* Cambridge, MA: Harvard University Press.

Stenmark, J. D., Thompson, V., & Cossey, R. (1986). *Family math.* Berkeley: University of California, Lawrence Hall of Science.

Sterrett, A. (Ed.). (1990). *Using writing to teach mathematics* (MAA Notes No. 16). Washington, DC: The Mathematical Association of America.

Tobias, S. (1993). *Overcoming math anxiety.* New York: Norton.

Tout, D., & Johnston, B. (1995). *Adult numeracy teaching: Making meaning in mathematics.* Melbourne: Adult Basic Education Resources and Information Service (ARIS), Language Australia.

Venezky, R. L., Wagner, D. A., & Ciliberti, B. S. (Eds.). (1990). *Towards defining literacy.* Newark, DE: International Reading Association.

Wedege, T. (1996, July). *Professional profile in mathematics of adults returning to education.* Paper presented at Working Group 18 (Adults Returning to Mathematics Education), Eighth International Congress on Mathematics Education, Seville, Spain.

Wickert, R. (1989). *No single measure: A survey of Australian adult literacy. Summary report.* Canberra, Australia: Department of Employment, Education and Training.

Willis, S. (Ed.). (1990). *Being numerate: What counts?* Melbourne, Australia: Australian Council for Educational Research.

Wolf, A. (1990). *Learning in context: Patterns of skills transfer and training implications.* Sheffield, UK: The Training Agency.

2

Numeracy, Mathematics, and Adult Learning

Diana Coben
University of Nottingham

This chapter brings together selected research findings and ideas about numeracy, mathematics, and adult learning. It examines the questions, What does it mean to be numerate? and What do we know about how adults learn mathematics?, and draws out some implications for adult numeracy practice and research.

INTRODUCTION

Questions about the meaning of numeracy and about learning processes of adults in this area have concerned me as an adult educator and numeracy practitioner over many years, and now as a researcher and academic in the field of adult education. My involvement with adult numeracy dates from the mid-1970s when I began working as an adult literacy tutor in London and found that many students asked for help with mathematics. Some had quite specific goals—anything from counting or telling the time to passing a mathematics entry test for a job or vocational training; others wanted to succeed more generally at a subject which they felt had defeated them at school. In response, my colleagues and I adapted the discussion-based methods used in literacy work, encouraging students to talk and write about the mathematics in their lives and gearing the lessons to their purposes and contexts. Some of us needed to brush up on our own mathematics and learned from our more numerate colleagues. As time went on, we produced teaching materials and ran training courses, sharing ideas with other tutors and students (Coben & Black, 1984). The numbers of those involved in what became known as "adult numeracy" grew.

As tutors we were responding to the demand from adult learners and our emphasis was on practical rather than theoretical questions. We were more concerned with questions of "how" than with questions of "why": how to teach a particular group of students, rather than why there were so many people seeking help or what the exact nature of their difficulties was. We agonized over the question "What is numeracy?" at the expense of asking "What is numeracy for?". Rather than addressing fundamental conceptual questions or questions of the purposes and contexts of numeracy in adults' lives, our discussions centered on how far we should go. Everyone agreed that numeracy included basic arithmetic, but was it okay to teach algebra? Or statistics?

Our answers to these questions inevitably varied: we went as far as our view of numeracy teaching and our own knowledge of mathematics would take us. However, the intensely practical focus of the work, together with our concern not to talk about numeracy in ways that would be inaccessible to students, led to a slight but nonetheless real prejudice in some quarters against theory. The result was that even though practice improved when tutors and students were in regular contact, elsewhere it languished; theoretical insights were largely unknown and research failed to develop.

The situation now is rather different. The publication of this book under the auspices of the National Center on Adult Literacy, and the establishment of the Adult Numeracy Network (ANN) in the United States, and of Adults Learning Maths (ALM), an international research forum founded in England in July 1994, shows that there is a new interest in linking theory, research, and practice in adult numeracy (Coben, O'Donoghue, & FitzSimons, forthcoming). The time is ripe to review what we know about numeracy, mathematics, and adult learning and to seek to apply that knowledge in order to enhance both practice and research. I begin by considering what it means to be numerate.

WHAT DOES IT MEAN TO BE NUMERATE?

The answer to this question, of course, depends on what you mean by numeracy—a notoriously slippery concept for which there are different definitions (see, e.g., Baker & Street, 1994; Evans, 1989; Gal, 1994; or Willis, 1990; Withnall, 1995a; see also chap. 1, this volume). However, the absence of an agreed definition of numeracy does not absolve writers on the subject from making clear the working definitions that inform their view; in fact, it makes it more necessary that the reader should know where the writer stands on this key issue. So what do I think it means to be numerate?

In my view, to be numerate means to be competent, confident, and comfortable with one's judgments on *whether* to use mathematics in a particular situation and if so, *what* mathematics to use, *how* to do it, what *degree of accuracy* is appropriate, and *what* the answer means in relation to the context. This perspective on the numerate adult is rather different from the one implied by the competency-based standards for the U.K. certificate in adult numeracy, *Numberpower*. These require the individual to demonstrate that he or she can, for example,

- handle cash or other financial transactions accurately using a calculator, cash register, interest tables, or other printed computation aids;
- keep records in numerical or graphical form;
- make and monitor schedules or budgets in order to plan the use of time or money;
- calculate lengths, areas, weights or volumes accurately using appropriate tools, for example, rulers, calculators, and so on.

These are skills in context, certainly, but what a limited set of contexts is implied, and what an impoverished view of adults' purposes in learning and practicing "basic skills"!

Willis' (1990) statement, "To be numerate is to function effectively mathematically in one's daily life, at home and at work" (p. vii), offers a broader and probably more commonly held conception of numeracy as primarily functional in relation to everyday life. Of course, individual adults' everyday lives are different, one from another, and Willis goes on to point out that the mathematical ideas and skills adults need in order to function effectively are changing and are likely to continue to change as society changes.

Another common conception of numeracy is that it consists in mastery of a list of decontextualized and largely computational skills, such as adding, subtracting, multiplying and dividing whole numbers, decimals and fractions, and so on. I see such lists as problematic for several reasons. First, they rest on the assumption that such skills are readily transferable, an assumption challenged by scholars such as Jean Lave (1988) and others (e.g., Nunes, Schliemann, & Carraher, 1993). Second, as Mathews (1991) points out, their implicit reference point is "mathematics as taught in schools, rather than the needs and experiences of people in and outside work" (p. 151). Third, an exclusive focus on skills (i.e., things people can do) says nothing about mathematical knowledge and understanding; furthermore, concentration on arithmetical computation means that other important aspects of mathematics, such as problem solving, spatiality, or algebra are not considered. The term *mathemacy* has been coined by Skovsmose (1992) to overcome this last difficulty and may well prove to be more useful than *numeracy* because of its broader focus.

Whichever term is used (and I use *numeracy* here because of its currency among practitioners), we are not discussing an absolute quality. There is no single point at which one can say that person is numerate (or mathemate) and that one is not. Instead, each individual has cognitive strengths in various areas—what Gardner (1983) calls "multiple intelligences," including logical-mathematical and spatial intelligences— and utilizes these strengths in different ways in different contexts for different purposes.

So, if the job of adult numeracy tutors is to help adults become more numerate, to become "competent, confident, and comfortable" with their mathematical judgments, then surely we need to understand as far as possible how adults learn, or fail to learn, mathematics.

WHAT DO WE KNOW ABOUT HOW ADULTS LEARN MATHEMATICS?

Here we encounter a problem immediately in that adult numeracy is generally under-researched, as shown by the fact that the otherwise excellent Basic Skills Agency (BSA) bibliography of basic skills research (BSA, 1994) devotes only 14 out of over 300 pages specifically to numeracy. Other recent sources giving a general overview of numeracy research include Gal (1993) and Thorstad (1991). The most recent of these, the BSA bibliography, groups research of relevance to adult/post-compulsory numeracy education as follows: mathematics in everyday life, mathematics and employment, numeracy policy, teaching of numeracy, and numeracy tutor training. It gives keywords by which readers may identify the areas covered in the research. Interestingly, the keyword "learning" appears in fewer than a third of the 44 sources listed under numeracy, and "learning theory" appears only once. The upshot is that theoretical questions about how adults learn mathematics do not appear to be very high on the adult numeracy research agenda.

But is there such a thing as an adult numeracy research agenda? Many of the sources cited in the BSA bibliography are of school-based research (K-12 in U.S. terms). Even though these may be very relevant to the adult numeracy field, they require some interpretation. Likewise, research in other disciplines such as cognitive psychology and the related fields of the philosophy of mathematics education (Ernest, 1994) and ethnomathematics (Frankenstein, 1987; Skovsmose, 1992) may also be relevant. As Iddo Gal (1994) remarks: "It is surprising that no attempts have been made so far to synthesize, interpret, replicate, or extend research of relevance to adult numeracy education that has been published by workers in other disciplines" (p. 15).

Agendas for research in adult numeracy, including interpretations of prior research, are only now beginning to emerge as numeracy practitioners and researchers come together and begin to explore their own needs and work, including practitioner research. UNESCO's first International Seminar on adult numeracy was held in Paris in 1993. In addition to the ANPN and ALM groups mentioned earlier, and to other groups such as the Adult Learners Special Interest Group of the Mathematics Education Research Group of Australasia (MERGA), who now hold annual meetings, the International Congress on Mathematics Education (ICME), for the first time held a working group on "Adults Returning to Study Mathematics" in its meeting in Spain in Summer 1996 (FitzSimon, 1997). As these developments gain momentum and lead to visible results, adult numeracy practitioners need to be resourceful in their reading of the literature.

So what do we know about how adults learn mathematics? Research on learning mathematics, from Piaget onward, has mainly focused on children's learning, whereas research on adult learning has mainly ignored the learning of mathematics, as exemplified by the fact that the index of Brookfield's (1986) otherwise comprehensive book on adult learning has no entries under mathematics or numeracy. Nevertheless, some things are known. I begin by looking briefly at general trends in research on learning mathematics.

Research on Learning Mathematics: From Behaviorism to Constructivism

There has been a major shift in ideas about learning mathematics over the past 25 years, as Kieran (1994) describes in a recent retrospective of research. She contends that whereas previously learning mathematics was equated with immediate recall, retention, and transfer, and understanding was equated with achievement in tests or the performance of tasks, now learning mathematics is regarded as "learning mathematics with understanding"—the notions of mathematical learning and understanding have merged. Skemp's (1971) distinction between instrumental understanding (i.e., following the rules without necessarily understanding the reasons behind them) and relational understanding (i.e., knowing what to do and why) may be seen as presaging the new emphasis on understanding.

Kieran argues that this shift reflects a change from behaviorist to constructivist research perspectives on learning mathematics. Whereas behaviorists look for evidence of learning in changes in behavior, constructivists see learning as understanding constructed by the learner and focus on ways in which the individual learner makes sense of mathematics (after Piaget), or, increasingly, see learning as an activity in which shared mathematical meanings are constructed socially (after Vygotsky).

Jaworski (1994) points out that current debates between "radical" and "social" constructivists (see Ernest, 1994) parallel the distinction between Piagetian and Vygotskian perspectives. In the radical constructivist view:

> Knowledge results from individual construction by modification of experience. Radical constructivism does not deny the existence of an objective reality, but it does say that we can never know what that reality is. We each know only what we have individually constructed. (p. 17)

By contrast, the social constructivist view emphasizes that the construction of knowledge is a social process of interaction and negotiation with others and with the environment, rather than a purely individual process.

Constructivism is a philosophical perspective on the origins of knowledge and the processes and outcomes of learning, rather than a pedagogy. Nevertheless, it is internationally recognized as having significant implications for mathematics education (see, e.g., Ernest, 1991, 1994), although these have yet to be worked out for adult numeracy. Certainly, constructivism seems to be in the ascendant in mathematics education in the United States and underlies the ideas presented in the recent NCTM *Standards* (see chap. 3). However, lest we get carried away with the idea that constructivism is sweeping all before it, the influence of behaviorism is still evident and influences thinking in some circles, as testified by the Numberpower standards in the United Kingdom listed earlier.

Absolutist Versus Fallibilistic Views of Mathematics

If knowledge, including mathematical knowledge, is constructed rather than given, where does this leave notions of mathematical truth? Here I find Lakatos's (1976) distinction between absolutist and fallibilistic views of mathematics valuable. In the absolutist view, mathematics is seen as a set of absolute truths determined by authority; doing mathematics means following the rules correctly (i.e., only Skemp's instrumental understanding is required). By contrast, in the fallibilist view, mathematics is seen as a human construct and therefore value laden, culturally determined, and open to revision; Skemp's relational understanding seems more appropriate to this view. The distinction between absolutist and fallibilistic views of mathematics in relation to mathematics education is discussed in depth in Ernest (1991).

My own conversion came as a result of the fallibilist view being pointed out to me (though not in so many words) some years before I became involved in adult numeracy. I had gone to a seminar on mathematics and art and was astonished to hear the lecturer say that π (pi) had been invented to solve an awkward problem in mathematics. It is a measure of how totally my schooling in mathematics had been embedded in an absolutist framework that this came as such a revelation. It had simply never occurred to me that mathematics was invented—until then, for me it just was. At that moment I began to see mathematics not as a fixed set of somewhat alien precepts but as a creation of the human mind.

One Mathematics or Several? Ethnomathematics and Mathematics in the Cultural Context

If mathematical meanings are constructed by the learner, as constructivists believe, and if a fallibilist view of mathematics is viable, does that mean that there is one mathematics common to all cultures, or is mathematics itself subject to cultural change and diversity?

At first sight the question of whether there is one mathematics or several may appear ridiculous: after all, $2 + 2 = 4$ wherever and whomever you are. However, this is to confuse the universality of truth of abstract mathematical ideas with the cultural basis of that knowledge, as Bishop (1991) points out. To illustrate this, he asks, for example, where the idea of negative number came from, or why the angles of a triangle add up to 180 degrees and not 100 or 150. Bishop has identified six fundamental mathematical activities that he contends are found in all cultures: counting, locating, measuring, designing, playing (e.g., games with scoring rules, or games of chance), and explaining. The forms these activities take, however, vary greatly between cultures and it is these variations that interest Bishop and other ethnomathematicians. Ethnomathematicians have accordingly been active in breaking down the North's ignorance of the South's mathematics, as illustrated, for example, in Zaslavsky's (1973) celebration of African mathematics, and in Joseph's (1990) powerful attack on eurocentrism in mathematics and mathematics teaching.

Recent summaries of studies of mathematics practices in a cultural context (i.e., outside the classroom) include, for example, Nunes, Schliemann, and Carraher (1993), Saxe (1991), and Maier (1991).

In his review of such studies from Brazil, Carraher (1991) shows that important mathematical concepts do appear to develop outside school without specific instruction. Also, "folk" or "street" mathematics is usually performed orally rather than written down, and relates directly to the context in which it is done, unlike formalized school mathematics. As Carraher points out, there are both benefits and costs in this: "The major virtue of self-invented or intuitive mathematics is its meaningfulness; its major liability consists in the limited conditions to which this knowledge may be useful or relevant" (p. 195).

Ethnomathematics is described by d'Ambrosio (1997) as "the underlying ground upon which we can develop curriculum in a relevant way" (p. 22). He argues for the incorporation of the results of anthropological findings into the curriculum as part of a global, holistic approach, rather than as isolated examples.

Transferability

Clearly, if adults are to become mathematically "competent, confident, and comfortable," it is important that they are not restricted to being so only in familiar contexts. As Strässer, Barr, Evans, and Wolf (1991) point out, the ability to cope with unfamiliar contexts marks the borderline between Skemp's instrumental and relational understanding. However, it would be wrong to assume that school mathematics is more readily transferable than street mathematics. In their research on proportionality, Nunes and her colleagues found the opposite to be true: "both flexibility and transfer were more clearly demonstrated for everyday practices than for the school-taught proportions algorithm" (Nunes et al., 1993, p. 126). It seems that although school mathematics deals in more mathematically powerful general procedures, these may be poorly learned and quickly forgotten. Street mathematics, by preserving meaning, enables the individual to keep track of the operations needed while calculating and to spot errors more easily.

Invisible Mathematics

In my own research with Thumpston on adults' mathematics life histories (Thumpston & Coben, 1995), we found that many people do not seem to transfer their mathematical knowledge from one context to another. We also found that people do not necessarily think of the mathematics they are doing successfully as mathematics, but rather as "common sense," reserving the word "mathematics" for the mathematics they cannot do. Consequently, success in mathematics remains unattainable while the mathematics they actually do remains invisible to them. Evans and Harris (1991) call this the "'no maths here, we're practical people' syndrome" (p. 202).

For example, a laborer who left school at age 13 having apparently been taught very little mathematics said, "if I had to come up and do something which involves mathematics . . . I wouldn't be able to cope with it." However, later in the same conversation he revealed that despite his poor education, he had no problems working out fairly complicated measurements when working as a saddler:

> When I was pretty young, I used to have to measure up when I was on the harness-making. I had to work out the size of the harness we were going to make. The horses varied in sizes and everything was different sizes. We used to use a measuring tape. The sizes were given by the fellow that was running the business. He used to say this was so-and-so. . . . For instance, a belly band might have to be 4 foot 6, a saddle 2 foot in diameter, or whatever [...]. Winkers, 16 inches by 8 inches. It depended on the horse. I know the horse is 14 hands or 8 hands high or whatever. We were cutting out leather and all that and you worked to the eight inches . . .

There are surely important lessons here, and in the research on street mathematics referred to earlier, for adult numeracy teachers, in that students may have failed at formal school mathematics rather than at mathematics per se; they may be doing mathematics with their hands and in their heads rather than on paper. By studying how people make sense of mathematics in a cultural (or everyday) context, mathematics teachers—including adult numeracy teachers—may be able to promote the transition from context-dependency to context-independence. To the extent that this is achieved, educators will be helping to make meaningful representations and procedures (as in street mathematics) more powerful, and powerful representations and procedures (as in school mathematics) more psychologically meaningful (Carraher, 1991).

So what are the implications of all these theories and findings for adult numeracy practice and future research? What do we make of these ideas in light of possible constraints on the contexts of teaching, in which the tutor may have little knowledge of the mathematics of cultures other than his or her own, and may never have heard of constructivism or fallibilist views of mathematics?

IMPLICATIONS FOR ADULT NUMERACY PRACTICE AND RESEARCH

Whether as consumers or producers of research, I believe an awareness of different theoretical frameworks, including constructivist ideas, can help adult numeracy practitioners and researchers to think about their work in a new light. For example, questions such as the place of arithmetic in numeracy teaching are recast through a constructivist lens in a way that cuts through the "how far should we go?" debates that bedeviled my early years in adult numeracy. The question of what mathematics is worth learning is addressed "in terms of conceptual developments rather than skills to be acquired" (Cobb, 1988, p. 90).

A constructivist perspective means we should try to understand how adults make sense of mathematics, rather than assume that telling them clearly how to do something is enough for learning to take place. In other words, knowledge cannot be given by the teacher, it can only be constructed by the learner within the inevitable constraints (helpful and supportive or otherwise) of the teaching situation.

The constructivist mathematics educator Paul Cobb (1988) offers a way of thinking about the kinds of communication (and miscommunication) that go on between teachers and students. Cobb contends that teachers' actions constrain students' constructions of new knowledge structures—in other words, what the teacher says and does inevitably affects how (or whether) the student makes sense of mathematics but not necessarily in the way the teacher intends.

Picture the scene: you have just explained something as clearly as you are able for the third time; the students are looking embarrassed and perplexed (or distressed, or angry, or bored), and you can see that the message is not getting through. Later you run through the incident again in your head and wonder what you could have said or done that would have helped them to understand. Cobb (1988) turns the problem on its head, contending,

> The task of accounting for successful instruction is not one of explaining how students take in and process information transmitted by the teacher. Instead, it is to explain how students actively con-

struct knowledge in ways that satisfy constraints inherent in instruction. (p. 88)

Skemp's (1971) distinction between relational understanding and instrumental understanding may also be useful to adult numeracy educators. In fact, relational understanding may be especially important for adult learners, who, as Brookfield (1986) points out, like their learning activities to be problem centered and meaningful to their life situation and prefer to work out what to do and why in a meaningful context. As Evans and Harris (1991) argue, any discussion of the practice of mathematics must confront the idea of context: "This implies a study of everyday life in situ (in context) and an emphasis on the 'lived experience' of people's lives" (p. 202).

The importance of the context and people's lives for the planning of learning experiences underlies Frankenstein's development of "critical mathematics," a perspective exemplified in her "text book" for adults, *Relearning Mathematics* (1989). Inspired by the ideas of the Brazilian adult educator, Paulo Freire, she sees her work as helping to bring about social change, while acknowledging that change may come in unexpected ways. Her approach involves focused work on issues directly of concern to the adults in her classes, using newspaper articles, statistics on current issues, cartoons, and other art forms. Her students keep "math journals" in which they reflect on their learning process and the meaning that mathematics has for them (Frankenstein, 1987; see BeMiller, 1987, for further ideas and information on using learning journals in adult mathematics learning).

Frankenstein (1989) aims to help adults overcome any anxiety they may feel and gain the confidence to realize that they can relearn mathematics, recognize that they already do intellectual work in mathematics, and take that intellectual work seriously. The mathematics practiced by Frankenstein's students is meaningful, related to recognizable, although not necessarily familiar contexts, and intellectually challenging. The emphasis on mathematics-in-context ensures that meaning is not sacrificed to technique—Frankenstein's students learn how to do mathematics without losing sight of the larger picture. The variety of sources used, including ostensibly nonmathematical sources such as cartoons, also ensures that the learning environment is as rich and stimulating as possible.

The importance of a varied approach is borne out by research with instructors and young unemployed people on government-sponsored Youth Training Schemes in the United Kingdom in a wide range of occupations and contexts (Wolf, Silver, & Kelson, 1990). Wolf and her colleagues found that the most effective way to facilitate skills transfer was to use a variety of contexts for teaching and a "whole task" approach rather than concentrate on small tasks and subroutines.

This idea as well as others related to cross-cultural aspects of mathematics teaching and learning are summarized in the "six pointers" that Harris (1989) gleaned from the literature and from her work with aboriginal children in Australia. These pointers are worth reproducing in full for their relevance to adult numeracy practitioners:

1. The aim of mathematics instruction is the communication, sharing, and development of mathematical meanings, not just the "top-down" passing on of the teacher's own knowledge of mathematics.
2. It is the learner who is the meaning maker. That is to say, individual pupils construct their own meanings according to their particular past experiences, their cultural world view, their assessment of what is expected of them, and their own interpretation of what has been said.
3. Because of the different world views of those who belong to the MT (mathematico-technological) culture and students who belong to a tradition-oriented non-Western culture, it is especially important for teachers in these situations to know all they can about the pupils and to be aware of the significance of these fundamental differences.
4. Knowledge of the students' first language is important, but this knowledge, and even use of the first language as a medium of instruction, is not in itself sufficient to bridge the gap, as basic philosophical differences may still block communication and prevent the learner from integrating new knowledge with existing knowledge.
5. Talk is important in the mathematics classroom. There needs to be pupil-to-pupil talk as well as the more usual teacher-to-pupil talk. Where it is difficult for teacher and pupils to understand each other because of wide differences in language and world view, then pupil-to-pupil talk is likely to be especially important.
6. A great deal more research is needed in Australia to explore the possibilities (and advisability) of creating truly bicultural mathematics programs within the existing bilingual education programs for Aboriginal children. (p. 92)

So where does this leave us? I believe adult numeracy teachers need to observe, listen to, and learn from their students and each other in order to appreciate learners' cognitive strategies in their informal and formal mathematical operations and encourage those that are effective. They should encourage students to explore mathematics at their own pace and in their own ways. They must tread carefully if they are to avoid adding to the mass of cultural pressures that tell oppressed indigenous and minority peoples everywhere that their cultures are inferior. They need to appreciate the richness and diversity of mathematical experience and expression in their own and their students' lives and recognize the mathematics implicit ("frozen" is Gerdes', 1986, term for it) in ostensibly nonmathematical activities and cultural artifacts (see Harris, 1987, and chap. 14, this volume, for further discussion of these issues).

Educators should also appreciate the role of noncognitive, affective factors in enhancing or hindering the learning process (see McLeod, 1994, and chaps. 1 and 5, this volume). For example, learners' may know a particular mathematical operation intellectually but may not have accepted the knowledge emotionally, with the result that, especially under stress, they are likely to revert to their emotional rather than their intellectual response. My own recent work on the numerical skills required of healthcare students (Coben & Atere-Roberts, 1996), highlights an area in which the need to perform under pressure of time and overwork and to achieve a very high degree of accuracy adds to the problems that we know many adults face.

Adult numeracy teachers need to be particularly sensitive to the possibility that an adult's formal experiences of mathematics learning have been unproductive and painful. It seems likely that if a learner's experience of mathematics has been one of failure, fear or indifference, this will be compounded by the absolutist view of mathematics and impede mathematical learning in adulthood. If, on the other hand, learners adopt the fallibilist view and seek to understand their past experience of mathematics and to retrieve the mathematics embedded and hence unrecognized in their experience, then perhaps the vicious circle of ignorance, fear, and avoidance of mathematics may be broken.

Links Between Research and Practice

The relationship between theory and practice is complex; it is not simply a case of reaching for a recipe for adult numeracy practice from a book of theories. This is partly because adult numeracy theory is relatively underdeveloped—the recipe book is not yet written—but even if it were available such an approach would still be inappropriate. As Jaworski (1994) notes, the relationship between theory and practice is not just one way: "Practice might be seen to manifest theory rather than exemplify it. Practice tends to be far more complex than theory predicts and a study of practice can valuably enhance theory" (p. 32).

Practitioners, of course, are ideally placed to undertake a study of practice—to become researchers. Undoubtedly we need more research on all aspects of adult numeracy, but that is not enough on its own: research and theory must connect with practice. I believe adult numeracy practitioners need a critical awareness of current debates and ideas. They also need to develop their own research skills in order to undertake classroom- or program-based research so that they can contribute to and help to shape those debates. In so doing they will begin to break down the barriers separating researchers from practitioners and become what Schön (1983) calls *reflective practitioners.*

The topics that adult numeracy practitioner-researchers might tackle are legion. For example, how do adults conceptualize mathematics and work through mathematical problems in context? What cognitive operations are involved, and how do cognitive and affective factors interact? How does adults' previous experience of mathematics affect their approach to learning? What is involved in the transfer of mathematical skills, knowledge, and understanding from one context to another?

We need to explore the part played by adults' purposes for learning in expediting or inhibiting mathematical learning, both formal and informal. We need to know better how to communicate across cultural, gender, and other socially constructed barriers and to explore the implications of such communication for our understanding of the nature of mathematical knowledge and learning. We also, as Frankenstein (1987) suggests, need "to investigate how differential treatment based on race and class interacts with mathematics 'anxiety' and avoidance" (p. 194).

Further research is needed in order to better understand the implications of adults' starting points, such as the effects of gender (including being male) and gendered ways of knowing, ageing and lifestyle, and the impact of transitions in adult life (e.g., becoming unemployed, becoming a parent) on mathematics learning. We should further explore the implicit mathematics hidden in nonmathematical activities and cultural artifacts. Research should also examine the interplay between mathematics learning and language, the effects of ageing on mathematics learning, and the broad area of mathematics learning in and for employment.

These questions and related ones have hardly been examined. A first step toward answering these questions can be found via surveys of existing research findings from relevant disciplines, for example, in the literature of anthropology, philosophy, and ethnomathematics, as well as in that of education, mathematics, and psychology.

Research into numeracy and adult learning is already being undertaken in a small way by individuals writing dissertations at the master's or doctoral levels; much of this work is described in *Dissertation*

Abstracts International. Academic institutions as well as agencies and organizations, need to lobby for research grants—too little funded research is currently underway on either side of the Atlantic. Some exceptions to this in the United Kingdom are research at the University of Essex on numeracy and active citizenship and Withnall's (1995b) research at the University of Lancaster on older adults' needs and usage of numerical skills in everyday life. Similar work in other countries on specific questions is cited elsewhere in this volume. There is undoubtedly also an important role for self-initiated research by groups of adult numeracy practitioners and learners working together on aspects of adult mathematics learning; this type of research is illustrated by chapter 11 and chapter 12 in this volume.

The key to progress in adult numeracy is research. As Lerman (1990) argues in relation to mathematics education generally, research is vital in enhancing the status of adult numeracy. With a stronger research base, including practitioner research, adult numeracy, in the broad sense in which I have discussed it here, can take its rightful place as an area of legitimate public, professional, and academic concern.

REFERENCES

Baker, D., & Street, B. (1994). Literacy and numeracy: Concepts and definitions. In T. Husen & E. A. Postlethwaite (Eds.), *Encyclopedia of education* (pp. 3453-3459). London: Pergamon Press.

Basic Skills Agency (BSA). (1994). *Basic skills research: Bibliography of research in adult literacy and basic skills, 1972-1992.* London: Author.

BeMiller, S. (1987). The mathematics workbook. In T. Fulwiler (Ed.), *The journal book* (pp. 359-366). Portsmouth, NH: Boynton/Cook.

Bishop, A. (1991). Mathematics education in its cultural context. In M. Harris (Ed.), *Schools, mathematics and work* (pp. 29-41). Bristol, PA: Falmer Press.

Brookfield, S. D. (1986). *Understanding and facilitating adult learning: A comprehensive analysis of principles and effective practices.* Buckingham, UK: Open University Press.

Carraher, D. (1991). Mathematics in and out of school: A selective review of studies from Brazil. In M. Harris (Ed.), *Schools, mathematics and work* (pp. 169-201). London: Falmer Press.

Cobb, P. (1988). The tension between theories of learning and instruction in mathematics education. *Educational Psychologist, 23*(2), 87-103.

Coben, D., & Black, S. (1984). *The numeracy pack.* London: Basic Skills Agency.

Coben, D., & Atere-Roberts, E. (1996). *Carefree calculations for healthcare students.* London: Macmillan.

Coben, D., O'Donoghue, J., & FitzSimons, G. E. (Eds.). (forthcoming). *Perspectives on adults learning mathematics: Research and practice.* Dordrecht, The Netherlands: Kluwer Academic Publishers.

d'Ambrosio, U. (1997). Ethnomathematics and its place in the history of mathematics. In A. B. Powell & M. Frankenstein (Eds.), *Ethnomathematics: Challenging eurocentrism in mathematics education* (pp. 13-24). Albany: State University of New York Press.

Ernest, P. (1991). *The philosophy of mathematics education.* Bristol, PA: Falmer Press.

Ernest, P. (Ed.). (1994). *Constructing mathematical knowledge: Epistemology and mathematical education.* Bristol, PA: Falmer Press.

Evans, J. (1989). The politics of numeracy. In P. Ernest (Ed.), *Mathematics teaching: The state of the art* (pp. 203-220). Philadelphia: Falmer Press.

Evans, J., & Harris, M. (1991). Theories of practice. In M. Harris (Ed.), *Schools, mathematics and work* (pp. 202-210). Bristol, PA: Falmer Press.

FitzSimons, G. E. (Ed.). (1998). *Adults returning to study mathematics: Papers from working group 18, 8th International Congress on Mathematics Education, ICME8.* Adelaide: Australian Association of Mathematics Teachers.

Frankenstein, M. (1989). *Relearning mathematics: A different third R—radical math.* London: Free Association Books.

Frankenstein, M. (1987). Critical mathematics education: An application of Paulo Freire's epistemology. In I. Shor (Ed.), *Freire for the classroom: A sourcebook for liberatory teaching* (pp. 180-210). New York: Heinemann.

Gal, I. (1994). Reflecting about the goals of adult numeracy education. In I. Gal & M. J. Schmitt (Eds.), *Proceedings: Conference on adult mathematical literacy* (pp. 19-24). Philadelphia: University of Pennsylvania, National Center on Adult Literacy.

Gal, I. (1993). *Issues and challenges in adult numeracy* (Report TR93-15). Philadelphia: University of Pennsylvania, National Center on Adult Literacy.

Gardner, H. (1983). *Frames of mind: The theory of multiple intelligences.* New York: Basic Books.

Gerdes, P. (1986). How to recognise hidden geometrical thinking: A contribution to the development of anthropological mathematics. *For the Learning of Mathematics, 6*(2), 10-12, 17.

Harris, M. (1987). An example of traditional women's work as a mathematics resource. *For the Learning of Mathematics, 7*(3), 26-28.

Harris, M. (Ed.). (1991). *Schools, mathematics and work*. Bristol, PA: Falmer Press.

Harris, P. (1989). Contexts for change in cross-cultural classrooms. In N. Ellerton & M. Clements (Eds.), *School mathematics: The challenge to change* (pp. 79-95). Geelong, Australia: Deakin University Press.

Jaworski, B. (1994). *Investigating mathematics teaching: A constructivist enquiry*. Bristol, PA: Falmer Press.

Joseph, G. G. (1990). *The crest of the peacock: Non-european roots of mathematics*. London: Penguin.

Kieran, C. (1994). Doing and seeing things differently: A 25-year retrospective of mathematics education research on learning. *Journal for Research in Mathematics Education, 25*(6), 583-607.

Lakatos, I. (1976). *Proofs and refutations: The logic of mathematical discovery*. Cambridge: Cambridge University Press.

Lave, J. (1988). *Cognition in practice*. Cambridge: Cambridge University Press.

Lerman, S. (1990). The role of research in the practice of mathematics education. *For the Learning of Mathematics, 10*(2), 25-28.

Maier, E. (1991). Folk mathematics. In M. Harris (Ed.), *Schools, mathematics and work* (pp. 62-66). Bristol, PA: Falmer Press.

Mathews, D. (1991). The role of number in work and training. In M. Harris (Ed.), *Schools, mathematics and work* (pp. 145-57). Bristol, PA: Falmer Press.

McLeod, D. B. (1994). Research on affect and mathematics learning in the JRME. *Journal for Research in Mathematics Education, 25*(6), 637-647.

Nunes, T., Schliemann, A. D., & Carraher, D. W. (1993). *Street mathematics and school mathematics*. Cambridge: Cambridge University Press.

Saxe, G. (1991). *Culture and cognitive development: Studies in mathematical understanding*. Hillsdale, NJ: Erlbaum.

Schön, D.A. (1983). *The reflective practitioner*. San Francisco: Jossey Bass.

Skemp, R.R. (1971). *The psychology of learning mathematics*. Harmondsworth: Penguin.

Skovsmose, O. (1992). Democratic competence and reflective knowing in mathematics. *For the Learning of Mathematics, 12*(2), 2-11.

Strässer, R., Barr, G., Evans, J., & Wolf, A. (1991). Skills versus understanding. In M. Harris (Ed.), *Schools, mathematics and work* (pp. 158-168). Bristol, PA: Falmer Press.

Thorstad, I. (1991). *Proceedings of a seminar on adult numeracy* (Working Paper No. 91-1). Colchester, UK: University of Essex, Department of Mathematics.

Thumpston, G., & Coben, D. (1995). Getting personal: Research into adults' maths life histories. In D. Coben (Ed.), *Proceedings of the inaugural conference of adults learning maths: A research forum* (pp. 30-33). London: University of London, Goldsmiths College.

Willis, S. (1990). *Being numerate: What counts?* Hawthorn, Victoria: The Australian Council for Educational Research (ACER).

Withnall, A. (1995a). Towards a definition of numeracy. In D. Coben (Ed.), *Proceedings of the inaugural conference of adults learning maths: A research forum* (pp. 11-17). London: University of London, Goldsmiths College.

Withnall, A. (1995b). *Older adults' needs and usage of numerical skills in everyday life.* Lancaster, UK: University of Lancaster, Department of Continuing Education.

Wolf, A., Silver, R., & Kelson, M. (1990). *Learning in context: Patterns of skill transfer and their training implications* (Research and Development Monograph No. 43). Sheffield, UK: Department of Employment.

Zaslavsky, C. (1973). *Africa counts: Number and pattern in African culture.* Westport: Lawrence Hill.

3

Building a Problem-Solving Environment for Teaching Mathematics

Peter Kloosterman
Indiana University

Bin Hassan Mohamad-Ali
MARA University of Technology

Lynda R. Wiest
University of Nevada, Reno

Despite the fact that much of the mathematics taught in adult numeracy programs is the same as that which has been traditionally taught in (K-12) schools, teachers in these two settings have often acted without much knowledge about each other's work or innovations. Clearly, there are some differences in the way that mathematics should be taught to these two populations. It is our feeling, however, that there is far more common ground between the two types of educational contexts than is commonly realized.

To help to build a better bridge between adult educators and traditional mathematics teachers, this chapter examines two major documents published by the National Council of Teachers of Mathematics (NCTM) and discusses their implications for adult numeracy education. Some of the questions considered are: Why were the NCTM documents developed? What is NCTM saying about the direction mathematics teaching should take and why should adult numeracy educators be concerned about that direction? What should be happening in a numeracy classroom? What is number sense and why is it important in numeracy programs? We feel that considering some of these issues will help adult educators see their programs in a different light and be better able to provide for the mathematics needs of adults as we enter the 21st century.

MATHEMATICS FOR THE 21ST CENTURY: WHAT SKILLS ARE REALLY NECESSARY?

Forty years ago, record keepers made a living doing pages and pages of computations. Those individuals needed to be fast and accurate, but how many people need to do computations with paper and pencil today? All indications are that the numbers are very small and decreasing. To take a simple example, what mathematics skills are needed to work in a fast food restaurant? Workers do not use paper-and-pencil computational skills. Rather, they must use classification skills to make sure that each type of hamburger is placed in the correct bin. They must use communication skills to decide who will make the french fries and how many batches will need to be made. They must use estimation skills to determine, from the size of the line at the counter and the normal amounts of business at different hours of the day, how many of each type of hamburger to make. In addition, they need to make common-sense decisions about what task should be attended to next.

Compare the mathematical skills required to work in a restaurant to the traditional activities and curriculum in middle school, high school, and adult education mathematics classrooms. How often do we expect classroom learners at any of these levels to make an estimate based on continually changing data, or allow them to discuss a problem with others before trying to come up with a solution? Do we often expect classroom learners to rely on their experience and common sense to answer a mathematics problem? Unfortunately, few learners in traditional classes have had much experience with these kinds of activities. Why is the gap between school mathematics and the mathematics of the workplace so great?

One explanation for the gap between mathematical experiences in school and the workplace is tradition. The curriculum has always been focused on computation, so that is what is expected in mathematics classrooms. There has also been the argument that students must learn the "basics" before they can be expected to master higher order skills such as estimation. There is some truth to this notion: an individual must have an understanding of addition and multiplication to make a reasonable estimate of the amount of food that should be prepared to meet the needs of customers standing in line at a fast food restaurant. On the other hand, individuals with limited computational skills are certainly capable of using common sense and coming up with creative ways of solving "real-life" problems. Far too often in our society, learners at all levels have been deprived of the opportunity to learn to solve problems because they have not mastered basic computational skills. Evidence from the National Assessment of Educational Progress (NAEP) indicates that many public school students only master basic skills when they practice those skills by applying them to solve more complex problems in later grades (Dossey, Mullis, Lindquist, & Chambers, 1988). We believe that the situation is much the same for adult learners—the best way for them to learn basic skills is through practicing those skills by applying them to realistic problems. Such problems also help learners acquire the divergent thinking skills that are so necessary in the workplace.

In this chapter we look at the recommendations of the National Council of Teachers of Mathematics (NCTM) for helping learners to become true mathematical problem solvers. We explain why these recommendations are important to adult educators; we provide glimpses into some of the practices that can allow learners to meet their own educational goals as well as become productive members of the workforce and of society. Along the way, we consider the constructivist philosophy of teaching, describe learning environments that promote mathematical thinking, and illustrate how the ideas embodied in the NCTM recommendations can be implemented in developing skills such as estimation and number sense.

NCTM and the Push for Change

The notion that mathematics instruction needs to change to meet the needs of individuals during the 21st century is hardly new. Of the reports on mathematics learning in schools that came out in the early 1980s, three prompted particular attention: *A Nation at Risk* (National Commission on Excellence, 1983), *Educating Americans for the 21st Century* (National Science Board, 1983), and *New Goals for Mathematical Sciences Education* (Conference Board of the Mathematical Sciences, 1983), all focused on the fact that calculators and computers were making the need for paper-and-pencil computational skills obsolete and that more focus was needed on teaching higher order thinking skills. Unfortunately, there was no real consensus on how to accomplish this aim. More importantly, the calls for change were coming from organizations that had little experience in teaching mathematics. Members of the NCTM, as professionals dedicated to the teaching of mathematics, felt that they should be involved in decisions on how to reform mathematics teaching in the United States.

Three documents published by the council provide the framework for its recommendations and speak to the needs of both traditional and nontraditional learners. The first, titled *Curriculum and Evaluation Standards for School Mathematics* and often referred to simply as the NCTM *Standards* (NCTM, 1989), outlines the mathematics content students should be expected to master. The second, *Professional Standards for Teaching Mathematics*, or just *Teaching Standards* (NCTM, 1991), provides examples of the types of classroom environments and instruction that are necessary for students to master the content outlined in the Curriculum Standards. The third, *Assessment Standards for School Mathematics* (NCTM, 1995), describes recommended assessment frameworks that can inform and support the teaching and learning processes envisioned in the first two documents.

A fourth document is currently being developed by the NCTM. Titled *Principles and Standards for School Mathematics*, the new document will include the themes and ideas from the first three while also providing more examples of how mathematics should be taught for the 21st century. In other words, while the NCTM documents described in this chapter will soon be reformatted and updated, drafts of the new document show that the message will remain essentially the same.

Each document represents over three years of work by literally hundreds of NCTM members involved in teaching, teacher training, curriculum development, and research in various settings and levels. The first document also formed the basis for the Massachusetts ABE Math Standards Project, which examined the NCTM standards for their applicability to adult education settings (see chaps. 11 and 12, this volume). The first two documents are discussed in this chapter, while a few of the ideas addressed by the third document are mentioned in chapter 15.

The NCTM Standards

The *Curriculum Standards* outline content students should master at three grade ranges: K-4, 5-8, and 9-12. Four key standards describe processes that are evident across the entire curriculum (regardless of the particular topic, be it arithmetic, algebra, geometry, statistics, and so forth). These are:

Problem solving. The first of the four general NCTM *Standards* involves "mathematics as problem solving." Specifically, when students finish high school they should be able to

> (a) use, with increasing confidence, problem-solving approaches to investigate and understand mathematical content, (b) apply integrated mathematical problem-solving strategies to solve problems from within and outside mathematics, (c) recognize and formulate problems from situations within and outside mathematics, and (d) apply the process of mathematical modeling to real-world problem situations. (NCTM, 1989, p. 137)

In short, this means that an individual is not mathematically literate unless she or he is able to think through and solve mathematics problems that are far more complex than the typical word problems in mathematics textbooks. Although the NCTM focus on being a problem solver sounds ambitious, the writers of the *Standards* felt that this goal can and must be met by all school graduates. Obviously, being able to solve complex, real-world problems is important for adult learners as well.

Communication. The second of the four K-12 standards outlined by NCTM involves mathematics as communication. Students at all levels are expected to express the thinking they use to solve a problem both verbally and in writing. Similarly, they are expected to understand and respond to the solution methods described by their peers.

Reasoning. The third standard, mathematics as reasoning, involves having students make and test conjectures, formulate counterexamples, and construct and follow logical arguments. In other words, learners need to be able to decide when mathematics is needed to solve a problem and then keep trying different ways of solving the problem until a reasonable solution is found.

Connections. The fourth K-12 standard involves mathematical connections. To really know and be able to apply mathematical reasoning, students must see connections between various mathematical ideas (e.g., adding a negative number gives the same result as subtracting that number when it's positive), and appreciate connections between mathematical ideas that come up in different school subjects and between in-school and out-of-school mathematics. To put this another way, learners who fail to see how the mathematics they are learning can be applied to solve problems in different areas are not likely to remember or use what they have learned. Much of the reason why students forget the mathematics they were taught in school is that they did not see connections between the rules they were learning, rules they had already learned, and the situations in which those rules could be applied.

Additional standards. Other important standards for mathematics learning vary by grade level and describe the mathematical knowledge and reasoning expected in specific areas, such as understanding place value, measurement, estimation, geometric reasoning, algebra, statistics, and trigonometry. These standards focus on specific mathematical content and clearly indicate that basic arithmetic and algebra skills that comprise much of the traditional mathematics curriculum are only two of several domains that mathematically literate individuals must master.

SCANS

At about the same time the original *Curriculum Standards* were published, a more general but complementary project was undertaken by Lynn Martin, the U.S. Secretary of Labor. The committee assigned to the task became known as the Secretary's Commission on Achieving Necessary Skills (SCANS) and was charged with (a) defining the skills needed for employment; (b) proposing acceptable levels of proficiency, and (c) suggesting effective ways to assess proficiency (SCANS, 1991). The commission, primarily comprised of individuals from the private sector, but also educators, looked at education from the perspective of what employers wanted job applicants to know. Meetings, surveys, and discussions with union and industry leaders led the commission to the conclusion that the world of work was changing. In the words of the Commission:

> A strong back, the willingness to work, and a high school diploma were once all that was needed to make a start in America. They are no longer. A well-developed mind, a passion to learn, and the ability to put knowledge to work are the new keys to the future of our young people, the success of our businesses, and the economic well being of the nation. (SCANS, 1991, p. 1)

This finding was hardly a surprise, and it clearly complemented the view of the work on which the NCTM documents were based. Recommendations from the report included teaching learners to be creative thinkers, decision makers, and problem solvers, and to visualize pictures, make graphs, and use mathematical reasoning when needed.

Although the authors of SCANS were less concerned about how to teach than what to teach, they concluded: "We believe . . . that the most effective way of learning skills is 'in context,' placing learning objectives within a real environment rather than insisting that students first learn in the abstract what they will be expected to apply" (1991, p. 19). To the extent that it is possible to simulate the work environment in schools, NCTM positions on how and what to teach in mathematics are entirely compatible with positions taken by the authors of SCANS. In short, recommendations from mathematics educators (NCTM) and from private industry (SCANS) have come to the same conclusion: Mathematics instruction must change so that learners are proficient at recognizing situations in which mathematical procedures are helpful and then choose or create the procedures necessary to solve the problem at hand.

Teaching Standards

As an introduction to the descriptions of a problem-solving environment that follow later in this chapter, it is appropriate to outline the major themes of NCTM's second document, the *Teaching Standards* (1991). This document was written as a guide to what instructors should do to assure that students master the mathematics content outlined in the *Curriculum Standards*. Themes of the *Teaching Standards* include:

1. *Worthwhile Mathematical Tasks.* Although some computational exercises are appropriate, NCTM recommends that learners explore a variety of more open-ended problems, particularly those that involve the application of mathematical ideas.
2. *Discourse.* Another theme involves discourse, which is defined as mathematical discussion between the teacher and students and, just as important, among students. From the SCANS perspective, workers who cannot share their mathematical ideas with others do not work well in teams and thus are limited in the sorts of jobs they can perform. (See chap. 10 for further discussion of processes and activities for developing discourse and mathematical communication skills).
3. *Tools.* NCTM recommends using a variety of "tools" to enhance exploration and problem solving in mathematics. Tools can be calculators and computers, but they can also be simple things such as drawings or charts that help learners and their peers to understand a problem and its solution. For adult learners, tools include newspaper articles that involve mathematics, technical manuals needed on the job, and machines for which the operator must perform mathematical calculations. Such tools provide opportunities for discourse

while at the same time help learners see the connections between school and out-of-school mathematics.

4. *Assessment.* The *Teaching Standards* include a number of recommendations about assessing student learning. Grading is not an issue in many adult education settings but continually improving instruction to meet the needs of the learners is. The *Assessment Standards for School Mathematics* (NCTM, 1995) states that (a) assessment should reflect the mathematics that is important for individuals to learn, (b) assessment should enhance learning, and (c) assessment procedures should allow students to demonstrate knowledge in a variety of ways. These principles emphasize the connection between assessment and good instruction and imply that assessment in numeracy classes should include a variety of mechanisms by which learners can prove to both instructors and themselves that they are able to solve challenging problems.

Why Must Numeracy Programs Change?

Although the NCTM has traditionally been an organization of teachers of elementary, secondary, and two-year college students, there is a growing thrust within the organization to do more to include adult learners and adult educators in its activities and initiatives. Although the push for change in mathematics instruction has been accelerated by national reports and public sentiment, it is also the result of instructors in all contexts of instruction, including adult education and workplace settings, asking themselves how they can most help their students.

Think of the individuals you teach. Can they tell you how the ideas they are learning apply to home or job situations? Do they see that logic and geometry are mathematical skills with everyday applications? Do they routinely estimate to make sure the answers they are getting are reasonable? Do they see mathematics as just a set of rules to be mechanistically applied, or as a collection of general techniques and ideas that can be pieced together as needed to solve complex home and workplace problems?

Traditional mathematics instruction has focused too much on drilling students to remember bits and pieces of computational routines. Teachers in seventh grade, think students learned nothing in sixth grade. Teachers in eighth grade think students learned nothing in seventh grade and so on. Teachers cover the material in classes but students do not remember it from week to week, much less from year to year.

Educators are finally coming to the realization that traditional instruction has not worked for many students and something else needs to be done. They realize that the skills learned from traditional instruction are only a subset of the ones needed on the job—even jobs as straightforward as working in a fast food restaurant. In short, teachers themselves feel that mathematics instruction needs to change. It was this feeling, as much or more than national reports, that caused teachers in NCTM to push for change.

In the next section, aspects of the learning environment envisioned by the NCTM as a vehicle for better mathematics instruction are discussed. This is done with the assumptions that (a) numeracy instructors will find many of NCTM's suggestions useful, and (b) more communication between instructors in K-12 school programs and numeracy programs can only strengthen the quality of both. In the environment outlined next, the teacher functions more as a coach than a lecturer and uses questions to guide learners to the solution of a problem rather than giving them the answer. The learners work collaboratively, and there is little fear of "making a mistake" because everyone is expected to make mistakes. It is this type of environment that is most likely to result in learners finally mastering the skills they will need in mathematics.

RETHINKING THE LEARNING ENVIRONMENT

Constructing Mathematical Knowledge

Embedded in the NCTM Standards documents is a philosophy of teaching commonly referred to as "constructivism." Although the documents never actually use the term, it is a main part of the foundation on which those documents are based. In brief, the constructivist philosophy is that learners must figure out concepts and ideas themselves (see Kloosterman & Gainey, 1993; chap. 2, this volume). As teachers, we can present ideas, but simply presenting ideas does not mean that students will learn. Learners must analyze and discuss new concepts, and see how they are related to what they already know to maximize the chances of retaining those new concepts. This notion is particularly appropriate in adult education in which most of the mathematics that learners are studying is mathematics that they have seen before but never mastered. Take, for example, a problem as simple as 0.4×0.6. Some learners are reluctant to accept 0.24 as an answer because they are used to thinking that "multiplication always makes bigger" and 0.24 is smaller than either 0.4 or 0.6. For these learners, the "knowledge" that multiplication always makes bigger gets in the way of understanding.

Previous knowledge can also be very beneficial to learning. House roofs are commonly constructed with a 3-12 pitch, a 4-12 pitch, or a 6-12 pitch. (A 3-12 pitch means that the roof raises 3 feet for every 12 horizontal feet.) A person with experience framing a roof should be able to draw on his or her knowledge of pitch to make sense of fractions, ratios, and possibly even linear functions and their graphs. All adults have significant knowledge of mathematics on which to build. Unfortunately, instructors sometimes fail to take advantage of that knowledge.

In brief, the goals of mathematics instruction are undergoing significant change. We have always wanted students to apply their mathematics skills, but in the past have spent much of our instruction in both public schools and in literacy programs drilling students on bits and pieces of mathematics. We now know that drill has limited value in teaching mathematics. In particular, drill does little for *long-term* retention of skills unless learners see how those skills can be directly related to something they already know or want to know. Furthermore, learners who have failed to master the fragments they have been taught have been shut off from learning to apply those fragments to solve more complex home and workplace problems.

NCTM, the SCANS Commission, and advocates of adult literacy agree that continuing to focus on bits and pieces of mathematics (or other topics) is a poor strategy for many learners, particularly adults. Instead, learners need to be actively involved in learning and in solving challenging problems. On some of those problems, learners are bound to fail on the first and often subsequent efforts. The successful individual, however, is one who perseveres to get the job done. As numeracy educators, we need to provide emotional support to those who fear mathematics but if we only give them easily solved mathematics problems, we are failing to prepare them to solve the real-world problems they will encounter. "Students will perform better and learn more in a caring environment in which they feel free to explore mathematical ideas, ask questions, discuss their ideas, and make mistakes" (NCTM, 1989, p. 69).

Although the above statement comes from the K-12 *Curriculum Standards*, it applies to learners at all levels. And even though it may appear easy to build such an environment in the classroom, getting learners involved in discussions that help them to understand mathematics is a very challenging task. The environment is one in which the instructor is a *facilitator* rather than a transmitter of knowledge and in which learners feel challenged but not unduly frustrated. Individual initiative is fostered and expected, and intellectual risk taking is welcomed.

Consider the following problem:

Suppose you have a square piece of cardboard that is 80 cm on each side. Your task is to cut and fold the cardboard to

make the largest possible shipping container out of this cardboard. How should you do it?

This problem requires the type of realistic problem solving
advocated by NCTM. The problem-solving process will include several
phases specific to this problem, but will also involve important processes that can enhance mathematical problem solving in general, as illustrated below:

- *Problem Formulation.* Decisions must be made about whether
  the container will need a top and whether it makes sense to
  try and join pieces of cardboard to make sides, or whether
  each side should be made from a single, and therefore much
  sturdier, section of the cardboard square.
- *Group Discussion.* Working in groups, learners might discuss
  problem formulation as well as chart a course of action. They
  may want to try to make their own containers, or scale models
  of containers, to get a feel for practical designs. Although
  some designs will probably be more appropriate than others,
  the learners themselves should come to this conclusion.
- *Patterns and Prediction.* An important skill for solving many
  problems is seeing numerical patterns. Learners should be
  encouraged to compare volume to size of each of the sides of
  the model containers to see how side length relates to volume.
  After constructing two or three models, learners should become
  proficient at calculating volume to determine the capacity of
  their models, as well as become able to predict volume of different sized containers without having to build those containers.
- *Generalization.* Once a reasonable solution to this problem is
  found, learners should be challenged to predict how to maximize volume when starting with different sized square and
  nonsquare pieces of cardboard.

This brief discussion aims to highlight that the envisioned learning environment should be a mathematical community where mathematical ideas are used in discussions of problems and issues. All learners must be encouraged to share their ideas and to come up with different ways of solving the same problem. Our experience has certainly
been that having variety in the way problems are defined and solved
means that almost everyone in a group contributes something toward
final problem solution.

Collaborative Learning

As alluded to in the previous example, collaborative learning mirrors the kind of work that occurs most often in a job setting, and it fosters the interpersonal skills advocated by NCTM and SCANS. Reasoning skills develop as group members discuss and debate mathematical ideas and propose problem solutions. Parents with children in school are likely to find, in marked contrast to mathematics textbooks of the past, that their children's texts contain activities that are to be completed in pairs or in small groups. Just as a better product results from a joint effort in industry, so does improved learning in educational settings.

Investigating mathematics with others exposes individuals to a broader range of ideas and ways of thinking. Learners see problems from different perspectives and acquire various strategies for approaching them. In describing "mathematically empowered" people, Schoenfeld (1992) says, "They are flexible thinkers with a broad repertoire of techniques and perspectives for dealing with novel problems and situations. They are analytical, both in thinking through issues themselves and in examining the arguments put forth by others" (p. 335). Discussing mathematical problems promotes communication about mathematics and use of the language of mathematics, important goals included in the *Teaching Standards*. Further, learners sometimes express ideas to each other in "unofficial" ways that teachers never think of, increasing their peers' comprehension and maximizing use of instructional time.

Instructor role in a group setting. A question often asked is what instructors are expected to do during group work. Like most teaching situations, the answer to this question varies with the personality of the instructor and the instructional setting. In most cases, however, the instructor should move around the room and interact with individuals in the groups. The instructor's role is to help the groups solve problems rather than to tell them how to solve problems. Instructors should listen to what learners are saying to their peers to get a better feeling for where groups might need guidance. They must be careful, however, to make the group do the work. We find that, given our years of experience with lecture style instruction, it is very difficult to step back and let our students do the work by themselves. In the workplace and in the public schools, however, this is happening more and more. Instructors need to let all learners, especially adults, learn to depend on themselves.

Instructional setting. Although we have used collaborative learning in both adult education and traditional classroom settings and have been delighted with the results, we feel obligated to point out that using collaborative learning is dependent on having a classroom setting geared for this type of instruction. A number of adult literacy programs, particularly on-the-job programs, function through an individual tutoring rather than a classroom basis. When numeracy educators are working only as tutors, group learning is not possible. For example, we observed one industry program in which learners were given workbooks and did most of the activities at home. They attended the literacy center only to discuss questions they had or to get a new workbook. The instructors in this center were able to pair some learners so they helped each other during breaks and lunch hours, but there were other learners who were doing the entire program on their own. Although working on one's own has many limitations, it is far better than not working on mathematics at all, and thus we want to make clear that collaborative learning is a preferred but not essential element of a productive learning environment.

Questioning

Whereas collaborative learning is dependent on having several learners together at the same time, questioning is an important part of teaching that must be incorporated in any instructional setting. According to Kroll and Miller (1993), "Effective teachers ask both recall questions and higher level questions—questions that call for explanation" (p. 70). Good questioning involves having students explain and justify ("prove" in any comprehensible manner) answers and solutions, both correct and incorrect. Explaining and justifying mathematical thinking clarifies the thinking and aids retention. Talking through their thinking often helps learners discover their own errors.

Effective questions probe and extend thinking, asking "what ifs." The following questions might be raised in discussion of the container problem mentioned earlier. "What if you had a piece of cardboard that was 1 m on each side? If the parts to be stored in the container were 15 cm wide and 10 cm long, would you change the way you designed the box?" In general, good questioning includes the following:

- *Well planned questions.* Instructors should plan appropriate questions before they present a problem. Although thinking of questions as you teach is important, most of us ask better questions if at least some of them are planned ahead of time.
- *Wait time.* Instructors should allow enough "wait time" after posing questions for learners to fully consider that which has been asked and to generate a well-thought-out response. It is

common for instructors to wait only one or two seconds after asking a question before posing another one. Learners cannot formulate an answer to a challenging question in this amount of time.

- *Follow-up questions.* Incorrect responses should be met with questions that stimulate thinking in a way that may lead to a successful solution.
- *Giving hints only as needed.* In responding to questions and requests for help from problem solvers (who have first sought help from group members), instructors should carefully pose questions or give hints that may inspire thinking in different and appropriate ways without, as previously noted, giving away answers.

In general, questions help an instructor monitor learner thinking and problem solving, but, more importantly, they guide individuals in becoming autonomous learners who consistently take responsibility for reviewing their own solutions and answers.

Sources for Worthwhile Tasks

An important condition for an effective learning environment is the availability of diverse instructional resources that can help teachers identify a variety of motivating and important situations or problems on which instruction will focus. The textbook and workbook, or the course instruction manual in a job setting, can be used as a springboard for learning, supplemented by other teaching aids. For example, a study of statistics could include not only work from the textbook or course manual, but from many other books, magazines, and printed materials as well. The institution of state-run lotteries in the last few years has provided a highly motivating context for the study of statistics. The advertising slogan for one of the Indiana lotteries is "You can't win if you don't play." Regrettably, many individuals think they might win if they do play. Although many players would rather not know just how small their chances of winning a major prize are, lotteries are something most adults have prior experience with and are interested in discussing.

The term *functional context* has been used to describe teaching skills in the context of their application. Another rich source of instructional tasks are the problems adult learners are facing on the job or at home. Industrial workers may be employed at production facilities where quality control is monitored using Statistical Process Control (SPC). Learners with some knowledge of SPC or other quality assurance processes, but also with familiarity with graphs and charts that appear in newspapers (including in their job sections), should be able to transfer that knowledge to lessons on data collection and graphing. Teaching data collection, graphing, and graph interpretation skills in the context of SPC or newspaper stories provides a functional context which is important because, in terms of the constructivist philosophy, it allows learners to build on the knowledge they have already acquired, whether formally or informally.

Affect

Thus far, most of our discussion has been on what to teach and how to teach when working with adult learners. An additional issue that is particularly important for adults is their emotions toward mathematics. In the United States there is a tendency to believe that learning mathematics is a question more of ability than effort (McLeod, 1992). Many adults thus believe that their inability to solve mathematics problems is a permanent state, and there is nothing they can do to remedy the situation. Moreover, these adults openly admit that they are not good at solving mathematical problems, even though they probably would not want to admit it if they could not read or write. Even teachers of mathematics sometimes give up on students whom they think do not have adequate ability. Obviously, learners who have this belief and, as a result, feel they cannot improve their mathematical ability by working hard are not likely to study to increase their problem-solving skills. These beliefs, both by the adult learners and some teachers of mathematics, will have to be addressed by teachers in a systematic way.

Confidence is one of the most important affective factors in understanding an individual's ability to solve problems and hence plays an important role in the learner's mathematics achievement (Hart & Walker, 1993). An individual who is having a "mental block" while trying to solve a mathematical problem is unlikely to be successful in the endeavor (McLeod, 1992). Instructors need to look for the reactions of learners during the process of solving mathematics problems and identify those who are affected by the loss of confidence. For these individuals, a different approach or a change of sequence of doing things may be helpful.

Overcoming other negative feelings about mathematics is equally challenging. Kloosterman and Gorman (1990) provide a number of suggestions for improving motivation and attitudes of mathematics learners. One suggestion is for teachers to communicate to students the belief that they can learn mathematics, especially problem-solving aspects of mathematics. Another suggestion involves communicating the difficulty of problems to students so they do not feel that failure to solve a difficult problem indicates they cannot do mathematics. In brief, *positive feelings about mathematics come from solving challenging mathematics problems.* (Additional detail on dealing with negative attitudes toward mathematics can be found in chap. 1 on affect, attitudes, and beliefs).

APPLYING THE NCTM STANDARDS: TWO EXAMPLES

We have discussed collaborative learning, questioning, and resources for instruction as they apply to the learning environment. We now present examples of how these principles apply to teaching two important but commonly neglected topics: estimation and number sense. These are skills that all adults have in some capacity but, because they have not been part of the traditional curriculum, they are not seen as the essential mathematical skills that they are. (See chap. 5 for additional discussion of these topics.)

Estimation

As noted in the introduction to this chapter, the infusion of calculators and computers into both the workplace and everyday environments has significantly reduced the need for performing paper-and-pencil computations. Estimation skills, however, are more important than ever because so many situations in real life call for producing an estimate (Sowder, 1992). The most common categories of estimation used in daily life are those of estimating results of computation (e.g., if I have $10, can I buy 12 quarts of oil at 89¢ per quart?) and estimating measures (e.g., what are the dimensions of this room I want to carpet?).

Comparison shopping. Consider the following situation. Suppose you are in a supermarket and need to buy spaghetti. It is available in an 8-ounce package that is being sold for $0.39 or in a 20-ounce package that is being sold for $1.19. Assuming you wanted the "best buy," how would you decide which package to purchase? Is this something you would do mentally, with paper and pencil, or with a calculator? One reasonable approach to this problem is to estimate the cost of 1 ounce of spaghetti for both the 8-ounce package and the 20-ounce package. Rounding to change the problem into a more mentally manageable form might result in an estimation of $1.20 divided by 20, or 6¢ per ounce, for the larger package. The price of the smaller package could be estimated at 5¢ per ounce by dividing $0.40 by 8. Another method of comparing prices would be to look for a common unit, such as 40 ounces. You could compare the prices of the two packages by multiplying the price per package by 5 and 2 respectively. An estimated price of $0.40 times 5 or $2.00 for the smaller package and $1.20 times 2 or $2.40 for the bigger package might result.

Discussion of estimates. We offered these two alternatives for estimating cost to demonstrate the range of ideas that good mathematical tasks can elicit. They are also intended to show how teaching students only to estimate by finding unit prices, the technique typically taught in textbooks, could impede discovery of the solution. If this task was presented to a class, the instructor's questions should be used to get students to see that both methods are reasonable estimations and to look for additional appropriate methods. Additional discussion should focus on accuracy of the estimates. Were they close enough to know for sure that the smaller size was a better buy? If not, what needs to be done to assure that the most cost-efficient product was purchased?

Validity of estimates. When estimating computations, as in the earlier example, validity of the estimation can be checked by comparison with the exact answer. In some situations, particularly those involving measurement, exact answers are not always possible. For example, it might be necessary to estimate the weight of a coal pile left outside the heating plant at the end of each year. In this exercise, groups would have to devise methods for determining the volume of the coal pile and the weight of a unit volume within the pile. In addition to learning estimation skills, learners would gain understanding of volume and density. The important aspect of such exercises is to be able to justify the strategy used to arrive at an answer and to evaluate its strengths and weaknesses.

Number Sense

Number sense has become one of the buzzwords of new mathematics curricula. In brief, number sense is both the ability and the desire to use numbers for a large variety of everyday needs (Sowder, 1992). One of the prerequisites to acquiring number sense is to develop intuition about large and small quantities and what it means to perform calculations with those quantities. Number sense includes, for example, making estimates of the bill in a restaurant or the time it will take to do several errands. It includes being able to make a reasonable calculation of the amount of fertilizer needed for a lawn or how to estimate the expense of running an air conditioner rather than several fans to cool one's home. People with number sense know how to determine the dosage of cough syrup that is appropriate for a child, or how to interpret a newspaper article on the odds of getting a certain form of cancer.

As noted earlier, one of the advantages of teaching adults is that they have numerous experiences on which to build. Because they have been forced to deal with quantities, many already have a key component of number sense: common sense. For example, many adults will be able to see that the package of 20 ounces of spaghetti is more than twice the size of the 8-ounce package and that they should pay a little more than twice the price of the smaller package to buy the larger one. Thus, using both an estimate and one's number sense, a learner should be able to say that the price of the larger package of spaghetti in the previous example should not exceed two and one-half times the price of the smaller package for the larger package to be a better deal.

National debt. Examples of large numbers that lead to the development of number sense are easily found in newspapers and news magazines. The national debt provides one possible context for a discussion of large numbers. When politicians begin talking in terms of billions or trillions of dollars, the numbers are so large that they become incomprehensible unless framed in an understandable context. In our instruction, we like to use the example of how much area is covered by a trillion $1 bills to help learners conceptualize what a trillion really means. (Note: terms such as billion or trillion have different numerical meanings in different countries.) By laying out 10 $1 bills in a 2 by 5 array, it is easy to see that 10 bills cover about one square foot. It follows that a million $1 bills cover about one million square feet. Because a square mile is 5,280 x 5,280 or about 28 million square feet, it will take $28 million to cover a square mile. Because there are about 36 groups of 28 million in a billion, this translates to about 36 square miles per billion dollars or 36,000 square miles for a trillion dollars. Thirty-six thousand is again too large a number to comprehend unless it is compared to something that is already known. Indiana has an area of about 36,000 square miles, so having learners relate a trillion to the number of bills necessary to cover Indiana helps to make a trillion comprehensible.

When we do the trillion dollar activity, we turn the problem of how much area a trillion dollar bills would cover over to learners and only give guidance, in the form of leading questions, when they are really stuck. If often takes 20 to 30 minutes to arrive at an estimate similar to the one just given, and much longer than if we just explained our solution. However, when a group finally arrives at a good estimate, the sense for large numbers that the individuals in the group have gained is well worth the time spent. We then expand on this theme by trying to get learners to provide examples of large numbers they are exposed to at home or on the job, using those numbers as a basis for other estimations and comparisons.

Converting measurements. Another example of number sense comes from work with the metric system. Given the number of metrically based products imported from other countries to the United States (which has traditionally used only the English or "standard" system of measurements), and the number of U.S. products being built for export, the metric system is increasingly becoming a part of life at home and in the workplace. An American citizen used to "miles per hour" or "Fahrenheit" needs not only to be able to appreciate what a speed limit of 100 kilometers per hour means when traveling in Canada, or what to expect when weather forecasters say the temperature is going to be 30 degrees Celsius, but to understand instructions and information coming from other countries. The activities we use when handling English-to-metric conversions often involve rounding, estimation, and number sense, rather than precise calculation. For example, we have learners examine metric measuring cups or metric wrenches to get used to small metric measures. We have them pay attention to the speed they are traveling, in kilometers per hour, on their way to and from class or work. We talk about easily converted measures such as the similarity in capacity of quart and liter bottles, and we expect learners to discuss situations in which they use metric measures and make measurement estimations. Overall, as in the earlier examples, we strive to develop learners' sensitivity to the relative magnitude of numbers and to their meaning in different contexts.

SUMMARY AND IMPLICATIONS

Our purpose in writing this chapter was to familiarize adult numeracy educators with the NCTM initiatives for change in mathematics instruction, with the hope that the bridge between K-12 education and adult education becomes stronger. Learners who have spent time in the workforce know what skills are needed on the job and thus are usually quite receptive to new ways of learning mathematics. Individuals without work experience have a tendency to think mathematics should be the same as it was when they were in school and thus often resist group learning and open-ended problems. As a numeracy educator, however, it is important to teach learners important mathematics, especially the ability to attack and solve challenging problems. The public schools are doing much more of this as are most industrial training programs. Learners need to recognize that time spent learning "pre-calculator" mathematics is time that could have been spent learning the estimation and number sense skills that calculators cannot perform.

To close this chapter, we provide a summary of our suggestions for applying NCTM principles in adult numeracy education settings (see also chap. 5 for further discussion of instructional principles).

- *Help learners see that mathematics is more than a set of rules to be memorized.* As noted, most adult learners have been subjected to years of "rule-oriented" instruction in mathematics. Thus numeracy educators must help them learn that rules are only one part of what it takes to successfully complete real-world mathematics problems.
- *Build on learners' previous knowledge of mathematics.* All adults use mathematics in some fashion, although many do not realize how frequently they use their mathematical skills. Numeracy educators need to talk with learners to find out where they are using mathematics skills and then help learners make connections between the mathematics concepts they understand and new ones they need to master.
- *Be a facilitator of learning rather than just a lecturer.* In a problem-solving environment, learners are encouraged to try a variety of solution methods and are not overly concerned when some of them fail to work. The instructor provides hints but expects learners to solve problems by themselves. Framing tasks in a functional context helps learners build on previous knowledge while also helping them see why the mathematical concept being taught is important.
- *Get learners to work together.* Collaborative learning can be very helpful because it both aids learning and simulates conditions of the home and workplace. At the very least, having pairs of individuals solve problems together helps them to see that discussing problems often leads to new insights for solution.
- *Ask complex questions.* Questioning is important because it helps learners at all levels to focus their thinking and to learn to "speak mathematics." It is also a good mechanism for helping learners make connections between new concepts and those already mastered.
- *Help learners to become confident in their ability to use mathematics.* Affect is a particularly important consideration for adults because many of them either do not like mathematics or feel that they cannot do it. Help learners to build confidence by seeing that complex problems often can be solved when attacked with vigor.

In brief, the time has come to bring mathematics instruction at all levels in line with the needs of citizens of the 21st century. Unfortunately, change is not easy and, in the words of the *Teaching Standards*, "Teaching is a complex practice and hence not reducible to recipes or prescriptions" (NCTM, 1991, p. 22). The encouraging news is that adults are certainly capable of becoming problem solvers, of thinking logically, and of viewing mathematics as a team sport. As teachers, we can and must help them to achieve these goals.

REFERENCES

Conference Board of the Mathematical Sciences (1983). *The mathematical sciences curriculum K-12: What is still fundamental and what is not.* Washington, DC: Author.

Dossey, J. A., Mullis, I. V. S., Lindquist, M. M., & Chambers, D. L. (1988). *The mathematics report card: Are we measuring up? Trends and achievement based on the 1986 national assessment* (Report No. 17-M-01). Princeton, NJ: Educational Testing Service

Hart, L. E., & Walker, J. (1993). The role of affect in teaching and learning mathematics. In D. T. Owens (Ed.), *Research ideas for the classroom: Middle grades mathematics* (pp. 22-38). New York: Macmillan.

Kloosterman, P., & Gainey, P. (1993). Students' thinking: Middle grades mathematics. In D. T. Owens (Ed.), *Research ideas for the classroom: Middle grades mathematics* (pp. 3-21). New York: Macmillan.

Kloosterman, P., & Gorman, J. (1990). Building motivation in the elementary mathematics classroom. *School Science and Mathematics, 90,* 375-382.

Kroll, D. L., & Miller, T. (1993). Insights from research on mathematical problem solving in the middle grades. In D. T. Owens (Ed.), *Research ideas for the classroom: Middle grades mathematics* (pp. 58-77). New York: Macmillan.

McLeod, D. B. (1992). Research on affect in mathematics education: A reconceptualization. In D. A. Grouws (Ed.), *Handbook of research on mathematics teaching and learning* (pp. 575-596). New York: Macmillan.

National Commission on Excellence in Education (1983). *A nation at risk: The imperative for educational reform.* Washington, DC: U.S. Government Printing Office.

National Council of Teachers of Mathematics. (NCTM). (1989). *Curriculum and evaluation standards for school mathematics.* Reston, VA: Author.

National Council of Teachers of Mathematics. (NCTM). (1991). *Professional standards for teaching mathematics.* Reston, VA: Author.

National Council of Teachers of Mathematics (NCTM). (1995). *Assessment standards for school mathematics.* Reston, VA: Author.

National Science Board (1983). *Educating Americans for the twenty-first century: A plan of action for improving mathematics, science and technol-*

ogy education for all American elementary and secondary students so that their achievement is the best in the world by 1995. Washington, DC: National Science Foundation.

Schoenfeld, A. H. (1992). Learning to think mathematically: Problem solving, metacognition, and sense making in mathematics. In D. A. Grouws (Ed.), *Handbook of research on mathematics teaching and learning* (pp. 334-370). New York: Macmillan.

Secretary's Commission on Achieving Necessary Skills. (1991). *What work requires of schools: A SCANS report for America 2000.* Washington, DC: U. S. Department of Labor.

Sowder, J. (1992). Estimation and number sense. In D. A. Grouws (Ed.), *Handbook of research on mathematics teaching and learning* (pp. 371-389). New York: Macmillan.

4

Preparing Adult Students to Be Better Decision Makers*

Robert Clemen
Duke University

Robin Gregory
Decision Research

Learning to make good choices is a key skill for adults. This chapter discusses how curricula for teaching numeracy skills can be expanded into an important domain that often is ignored by adult (and K-12) educators: the practice of decision making. The chapter suggests ways in which often-neglected concepts and skills related to probability and sta-

*We are grateful to the participants of the "Literacy and Work Roundtable" held in Portland, OR during November, 1991, at which an early version of this chapter was presented. We also acknowledge the outstanding contributions by the editor, Iddo Gal, who provided extensive comments that were critical for converting our original paper into an integral chapter in this volume. Funding for our work was provided by Grant No. MDR-9154382 from the Instructional Materials Development Program, National Science Foundation. Any opinions, findings, and conclusions or recommendations expressed here are those of the authors and do not necessarily reflect the views of the National Science Foundation.

tistics can be developed in the context of curricula and case studies in decision making.

A WORLD OF DECISIONS

Save endangered species. Use paper instead of plastic. Buy car insurance. Ban chlorofluorocarbons (CFC). Regulate electromagnetic radiation. Reduce logging. Eat more fiber. Don't drink and drive. Close nuclear power plants. Produce less trash. Exercise regularly to reduce heart attacks. Save more for a rainy day. Smoking can be hazardous to your (baby's) health. Floss your teeth every day. Beware of AIDS.

Such slogans or expectations are unavoidable parts of today's social and economic landscape. Behind every slogan stand numerous individuals or organizations who, thoughtfully or not, have made their choices and taken a stand. And behind them lie a myriad of scientific or pseudo-scientific studies that are cited as support for the exhortation. Every slogan, and the life situations to which it relates, demands a decision. For example:

- Should I sign a petition to oppose the construction of a trash incinerator in my neighborhood?
- Should I join the demonstration against the closing of the neighborhood school?
- Should my family start recycling?
- Should I try to buy instead of rent an apartment?
- Should the United States participate in multinational efforts to save the ozone layer? Does it mean I have to throw out my old refrigerator?
- Should I (or my partner) use a condom?
- Should my kids be buckled with seat belts at all times?

These are some of the many choices we need to make. Along with broad awareness of scientific principles, facts, and methods, today's adults need preparation in decision-making skills tailored to the hard problems they face. Such skills can be taught to the nation's adult learners as part of a broad-based program in numeracy that also teaches the fundamentals of probability and statistics (though not in the alienating way these subjects have been traditionally taught). Bringing decision making into conventional subject areas, particularly when combined with authentic- and cooperative-learning methods, can capture student interest and teach adults important individual and group decision-making skills.

The study of decision making provides many useful concepts and tools for addressing difficult choices (Clemen, 1996). Over the past three decades, the methods of the decision sciences have become more accessible and now are widely applied in a variety of situations, ranging from personal decisions to corporate strategy and governmental policy. Although the use of some decision tools requires a sophisticated understanding of college mathematics, it is our belief that all individuals can understand the key elements of a framework for improved decision making. Further, we maintain that by teaching this framework to all adult learners, U.S. citizens will be far better equipped to address the difficult personal, communal, and professional choices they face today and will face in the years ahead.

DECISION MAKING, CRITICAL THINKING, AND HIGHER ORDER THINKING SKILLS

It is a simplification, albeit a correct one, to say that we want adult learners to be able to think critically and carefully about the decisions they have to make. In many ways this desire reflects the calls for "critical thinking," or "thoughtfulness," or "higher order thinking" that often are sounded by educators and politicians. All such concepts imply more than rote memorization, tiresome drills, and rehearsals of fact. The intent is for the adult student to do something more: to seek additional information when it is needed, recognize inconsistencies in a problem formulation, evaluate the truth of claims made in a statement or text, and combine information and techniques to cope effectively and responsibly with new situations.

Several of these functions are described by the term "critical thinking," although this is perhaps one of the most abused concepts in the current lexicon of educators. Ennis (1962) gave this term an initial boost by specifying 12 thought activities or characteristics that jointly define critical thinking. Since then, the concept has been used to indicate virtually any aspect of thinking beyond rote memorization and recall. Some researchers have been very specific about what "critical thinking" means; for example, it is "the assessing of the authenticity, accuracy and/or worth of knowledge claims and arguments"(Beyer, 1985, p. 272). However, these skills are only a part of careful decision making, which might include questions such as:

- How can the uncertainty in presumably scientific claims be represented and evaluated?
- What makes a measurement or estimate (e.g., of a probability that something will happen) reliable?

- How can the quality of information I may want to use for a decision be evaluated and improved?
- How do I resolve inconsistencies between different claims about the same issue?

Facing such critical questions and assessments is crucial as a decision maker marshals the information that pertains to a hard decision (Gregory, 1991).

Two other terms with relevance to a description of aspects of decision skills are "higher order thinking" and "thoughtfulness." As defined by Newmann (1990), higher order thinking "challenges the student to interpret, analyze, or manipulate information, because a question to be answered or a problem to be solved cannot be resolved through the routine application of previously learned knowledge" (p. 44). Although this concept seems to mirror the thought processes required for careful decision making, it is more general and does not explicitly address the question of decision techniques or include the role of personal values.

"Thoughtfulness" is not an actual thinking skill or activity but rather a general disposition. In paraphrasing the literature on thoughtfulness, Newmann (1990) describes thoughtfulness in terms of four traits: "a persistent desire that claims be supported by reasons (and that the reasons themselves be scrutinized), . . . a tendency to be reflective, . . . a curiosity to explore new questions, and . . . flexibility to entertain alternative and original solutions to problems" (p. 47)."

Newmann is particularly worried by the general absence of thoughtfulness in U.S. classrooms. The relationship between thoughtfulness and decision making is clear enough because careful consideration of issues is more likely to be pursued by an individual who is critical, reflective, and curious. Developing such dispositions is a tricky business, but many of the tools developed by decision scientists are designed to address the operational needs of a "thoughtful" approach to solving problems and making choices.

ELEMENTS OF DECISION MAKING

Decision skills can help the adult student to make decisions and to understand the impact of decisions on self, friends, neighbors, community, and society as a whole. The skills are particularly helpful for adult students because they often must consider personal values and tradeoffs in a practical and responsible manner. The structure of most problems faced by adults, in fact, emphasizes the practical aspects of choice: What matters to me? What options do I have? How will my choice affect what happens to me and others? How do short- and long-range consequences differ? Careful use of decision-making tools can help adult students discover insights about themselves and their world in addressing these fundamental questions.

At this point some readers may be asking, What do these lofty goals have to do with teaching numeracy? As explained later, there are certain relatively simple quantitative activities involved in handling decisions, as well as some simple statistical ideas. So, if anything, the ideas offered here provide a new and realistic context in which to practice some mathematical skills. Yet, notice that the main goal of this chapter is to help us set our sights high on an important curricular (and functional) goal, one that adult educators often list when discussing literacy in its broadest sense, but one that is only infrequently addressed in the context of teaching and learning numeracy skills. Working to improve decision-making skills enables us to develop critical and reflective thinking, thus achieving goals common to both numeracy and literacy education.

Although there are many ways to look at decisions, one that is especially useful for helping individuals to deal with difficult choices is known as "decision analysis," a discipline that springs from such diverse scientific fields as economics, statistics, psychology, and engineering. As defined by Keeney (1982), an originator of the approach, decision analysis is "a formalization of common sense for decision problems which are too complex for informal use of common sense" (p. 806).

In principle, most adults already have the rudiments of most decision skills, acquired through prior education, trial and error, life experience, and observation of others. Our educational mission is to help adult students further develop these skills. Yet, even if many adults never encounter decision situations so complex so as to require the application of formal decision analysis as described later, it is the process of learning to approach decisions in such a formal way that promises to help in developing thoughtfulness and reflectivity.

The methods of decision analysis help an individual to disentangle the major issues, understand more fully his or her values, identify relationships among elements of the problem, understand the key sources of uncertainty involved, and evaluate the quality of information pertaining to important outcomes. As with other decision-making aids, decision analysis only makes recommendations based on the available information; it cannot ensure a good choice or guarantee a good outcome. Likewise, not all decisions are important enough to deserve extensive analytic thought, although good decision-making skills include the ability to recognize when in-depth analysis is required (see the discussion in chap. 1 about what is entailed in being numerate). Even with these considerations, decision analysis is an important tool to help people think hard and systematically about the important choices they face.

Three key aspects that underlie decision analysis are shown in Figure 4.1 and discussed later: identifying important decision issues, understanding sources of uncertainty, and recognizing critical tradeoffs. The figure shows how these three areas are related; a decision maker must understand all three as a prerequisite to making a decision. And although it seems reasonable to begin by understanding important decision issues, subsequent consideration of uncertainty and tradeoffs may lead the decision maker to reconsider and redefine the main issues.

Identifying and Structuring Important Decision Issues

The first step in handling decision situations involves the adoption of a "decision perspective" in order to make sure all key aspects of a situation can be addressed. As an example, take the problem of deciding whether or not to buckle (fasten) one's seat belt. If a person adopts the here-and-now perspective of deciding each time whether to buckle up or not, it is easy to justify not buckling; the chance of an accident during a particular trip to a particular destination on a specific occasion is very small and, after all, buckling up entails some discomfort. However, if one adopts a long-range perspective, additional issues emerge. Pertinent information may include facts such as: (a) a person is quite likely (about one chance in three, in the United States) to be involved in a serious accident sometime during his or her lifetime; (b) as shown by many studies, seat-belt use greatly diminishes the chance of serious injury and subsequent decline in the quality of one's life; and (c) a serious injury is much worse than the small discomfort of wearing a seat belt.

The seat-belt decision can be construed not as whether to buckle up this time, but whether to adopt a policy (i.e., develop a habit) of using a seat belt consistently. From this perspective, the one-time small chance and the marginal discomfort are less relevant. The same can be said of other decisions people face, such as using drugs or condoms or smoking cigarettes, though in these cases the range of short-term and long-term issues to be considered may be greater.

In the process of structuring a decision, the decision maker has to break down and clarify the factors or questions that make a decision a difficult one: Are there value tradeoffs? Are there unknown risks, unknown costs of actions, or imperfect information about the likelihood of certain outcomes? How much is known about the ways aspects of the decision situation relate to one another? What is involved in obtaining guidance or more information regarding the various aspects of the decision? These and related issues are taken up in the next steps.

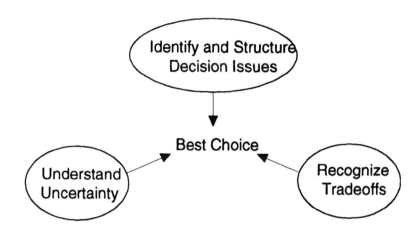

Figure 4.1. Three key aspects of good decision making

Understanding Sources of Uncertainty

What facts does a person know and what more can he or she find out about the decision situation? In some cases people have incomplete information; more time and research could uncover more relevant data or factors. For example, imagine that a person who moves to a new city needs to rent an apartment; he or she wonders if a certain apartment is overpriced compared to other apartments. Some research can yield reasonable information about the range of prices in the city (e.g., by scanning local newspapers or calling some rental agencies).

In other situations, however, we cannot hope for perfect knowledge because we cannot foresee the future. For example, when an adult has to decide what job training program to register for, how is he or she to assess how many jobs in a certain city will be available 10 years from now in the auto industry as opposed to the chemical industry? Or, take a more common example: What will the weather be like for next Saturday's picnic? No amount of research will tell us what is going to happen for sure, although what we learn or happen to know can help us to make more informed decisions (e.g., an August picnic will give most U.S. residents a higher probability of good weather than a date early in February).

Understanding the key sources of uncertainty can help to clarify the value of additional information and the urgency of the need to obtain such additional information: Do we know enough to make a decision now, or would it be prudent to learn more before making a decision? In general, the more consequential or irreversible the outcomes of the decision, and the less expensive (in terms of time, money, or stress) it is to gather information, the more sense it makes to learn more before making a choice.

Finally, understanding the sources of uncertainty also illuminates the nature of various risks that we face. It also highlights the need to consider from whom or where information about the likelihood of such risks be obtained, and the need to attend to the quality, completeness, and meaning of information that decision makers can obtain or generate. What does it really mean for a specific group (one's community) to be exposed to or reduce a particular risk, such as the impact of the construction of a new highway overpass on reduction in the expected number of traffic injuries or fatalities in their neighborhood? How can a community estimate the probabilities involved, and how credible or complete will the available information be, given any hidden or explicit agendas of the actors involved? Or, what does it mean for a specific individual (your best friend) to be exposed to a particular risk, such as the chance of serious side effects after going through a surgery, contracting AIDS if a condom is not used, giving birth to a baby with some congenital birth defects due to the medical history of the mother, or problems with a child's intellectual development due to lack of attention to proper nutrition?

Recognizing Critical Tradeoffs

Understanding tradeoffs means realizing that many of our objectives may conflict. How do multiple personal (and social) objectives relate to the decision situation and possible outcomes? What tradeoffs are involved? What does the decision maker (or society) want to accomplish? In what ways do the alternatives or possible outcomes either help to achieve or frustrate the decision maker's objectives?

Decision-making methods can help individuals clarify their values and objectives in order to make hard decisions in a coherent and defensible way. For example, should one move into Apartment A, which is expensive but located near a good school one's kids can attend, or should a person consider apartment B, which is much less expensive but will force his or her kids to attend a lower quality school? Or, should one look for a job (or attend school) in city A, where many of one's favorite relatives live, or should you move to city B, where salaries are higher? Both proximity to one's extended family and future earnings are important, but it looks as though more of one will mean less of the other.

How does one make tradeoffs between family values and future earnings? And where does a person draw the line: How much income, and how much time with one's relatives, does one really want?

DECISION MAKING CURRICULA AND ADULT LEARNERS

Formal courses in decision making typically are included in the syllabus that is designed for graduate and advanced undergraduate students in colleges and universities. Our experience as college instructors, however, suggests that no advanced training is required to apply the fundamental principles and basic tools of decision making. A more critical element is the willingness to think through a problem carefully and to look at it from different perspectives. Fortunately, this decision-structuring phase of decision making can be both intuitively appealing and fun and so can be a part of the classroom activities of adult numeracy learners.

This does not mean that casual thought automatically leads to a good decision. In fact, even though the tools may be basic, the thinking and analysis required can be challenging. The basic procedure is to construct a comprehensive picture, or model, of a decision situation (Plous, 1993). Modeling requires careful thought and a willingness to try various approaches as one develops understanding and insight into the particular situation. This modeling process is not one that can be reduced to a set procedure or applied blindly. Although following some specific steps can assist in the modeling process, those steps rely on the deliberation, reflection, and introspection of the decision maker.

The basic decision framework that underlies the phases depicted in Figure 4.1—constructing a model by identifying the critical elements of a situation, understanding uncertainty, and clarifying preferences and tradeoffs—can be used by virtually anyone. For example, consider a person deciding whether to purchase theft insurance on the contents of a new apartment. The most important issue (assuming no one is home) is the possible financial loss if a burglary occurs. To understand the problem, however, the individual must come to grips with the idea of risk; the insurance provides protection against an unfortunate occurrence that might happen, but a burglary is not guaranteed to occur. What is such a risk worth in terms of the relevant costs? Considering only the risk aspect, this decision could be pictured as in Figure 4.2.

The decision tree in Figure 4.2 is admittedly a highly simplified picture of a more complex decision. Developing the skill to depict a decision opportunity as a "decision tree," however, can be extremely useful for (a) representing the possible states of the world, (b) distinguishing between what is within our control and what is open to chance, and (c)

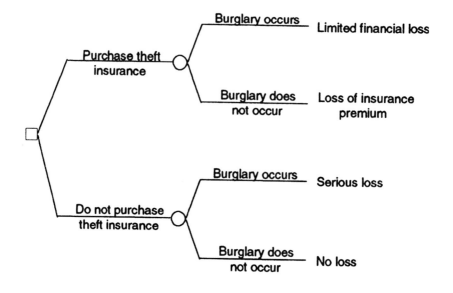

Figure 4.2. A decision tree for the theft-insurance purchase decision

initiating a discussion of uncertain events and their implications. Decision trees and similar techniques also can stimulate discussions about additional factors (e.g., area of city, value of contents, the cost of security systems) that might, in this case, influence the decision to either purchase insurance or accept a large possible future loss.

Each of the branches on the right side of the decision tree is characterized by uncertainty. The probability that an event will occur can be described in several ways. The most common approach is to use verbal descriptions, for example, referring to an event as "highly likely" or "reasonably probable" or "very unlikely." Such verbal descriptions, although useful in casual conversation, may lead to misunderstanding or confusion because different people may interpret the terms in different ways. For example, one person may think that an event with "low odds" will happen only about 5 times out of each 100 events (i.e., 5% of the time, or a probability of .05) whereas another person might think that it could happen as often as 1 time in 3 (i.e., 33% of the time, or a probability of .33).

Differences in the interpretation of verbal probability estimates can be significant. In such cases, decision scientists recommend that quantitative expressions of uncertainty be used, reflecting numbers between 0 (for an event that never occurs) and 1 (for a certain event). The rules of probability, as typically taught in a statistics module or course, are well-defined and permit the adult student to evaluate carefully the uncertain aspects of a pending choice. Placing probability and statistics in the context of decision and risk situations can stimulate students' interest in the concepts because their application is both practical and transparent.

Developing a decision-making curriculum. By asking adult students to address difficult personal and social decisions, adult educators can take their students beyond learning scientific and mathematical facts. They must also be exposed to, and encouraged to think about, important personal and social issues that have arisen from recent advances in mathematics, science, and technology. They should be asked to consider their personal values in light of this progress and to examine the impacts of decisions on themselves, their families and friends, or society in general. Consider, for example, the following questions:

- What are the consequences of new developments in biotechnology or new laws to protect the environment? In what ways are these consequences good or bad?
- What are the tradeoffs involved in promoting the use of generic over brand-name drugs?
- What consequences are associated with using nuclear rather than oil-based or coal-based electric power plants?
- What are the tradeoffs involved in cleaning up hazardous-waste sites?
- What are the pros and cons of allowing a natural area near a city to be used for a housing development?

In thinking about such questions, we want students to consider the limits of the decision maker's or society's knowledge in the pertinent areas and how that knowledge might be extended. Will new benefits or new risks be shared by society as result of the actions that might be taken? What more should be known before a decision is made? In a project funded by the National Science Foundation, we created a number of curriculum modules that allow a teacher to focus on various aspects of decision making (Gregory, Clemen, Satterfield, & Stone, 1996). (Other examples of decision-making curricula efforts include Baron & Brown [1991] and Beyth-Marom, Dekel, Gombo, & Shaked [1985]).

The goal of our project is to enhance student learning of science and mathematics concepts by focusing on decision skills: clarifying objectives, forming hypotheses, interpreting the outcomes of actions, creating alternatives, and evaluating the quality of information. Students also learn to make critical decisions about their work process: should they work alone or as part of a group and, if the latter, how should responsibilities (and blame or praise) be shared? When students—especially adult students—learn that many concepts from the mathematics or science classroom may be applied in important decision-making situations and extended to everyday life (e.g., with regard to family decisions), their interest in school-based numeracy tasks may increase. In addition, teachers find that training in decision making often helps students to stay on task, organize presentations more effectively, and work better as members of small groups.

IMPLICATIONS

The more we learn about ourselves and about our community, as well as about science, statistics, technology, and our planet, the more questions we face. These questions are difficult in a number of important ways. The fast pace of society often requires that decisions be made quickly, with imperfect information and with little knowledge of possible future technologies. Health and environmental risks require us to consider threats to individuals, to society as a whole, and to the environment. Difficult tradeoffs must be made, both at the personal level and at the communal and societal level, and these tradeoffs often reflect altogether new considerations. The 15-year-old who now worries about bicycle helmets will, in 10 years, be worried about biodiversity and the integrity of the ozone layer and, 20 years later, health risks from heart attacks or skin cancers. The adolescent now concerned about "safe sex" may, in 20 years, be worried about global overpopulation. Perhaps there will soon be a whole new set of decisions to be made. Twenty years ago, how many of us were concerned about greenhouse gasses or AIDS?

Science and technology present society with important issues and choices at a mind-boggling pace. As the pace intensifies, the gap grows between the scientifically literate and the decision makers—all of us—who must confront these important issues and make choices. Many issues require personal decisions that may in turn presume substantial knowledge: Should I use this pesticide in my kitchen or garden? What are the short-term benefits to me? What are the long-term risks? What evidence is there and where does it come from? Should I believe the source? Is a substitute product available? Often the same issue surfaces at a societal level: Should the use of this pesticide be regulated? What are the overall risks of the substance to society as compared to its benefits?

The gap in scientific literacy and numeracy is well known, and numerous calls and efforts have been directed at narrowing it. This chapter reminds the reader that the same gap also exists in terms of decision-making skills. We expect adults to be good decision makers and to find their way in a complex and multifaceted personal and professional world. Yet relatively little attention has been directed toward improvements in decision-making training for adults. As a result, most people remain in the dark regarding even the rudiments of systematic decision making, unprepared for upcoming choices, distrustful of most technical analyses, unsure how to function or negotiate over alternative points of view in a group setting, and largely unable to distinguish between useful and irrelevant information.

Most textbooks aimed at adult numeracy learners present basic concepts and terms related to probability in a mechanistic fashion and require students to solve simplistic questions involving fully known probabilities (e.g., results of coin flipping, die rolling, etc.). Likewise, students learn how to construct bar graphs or pie charts and how to "lift" or read simple data in such graphical displays, but not for the purpose of finding data to inform decisions. In contrast, this chapter suggests the need for a new agenda for teaching students statistics and probability, one involving development of students' ability to make use of the ideas behind the numbers and procedures they learn in probability and statistics for the purpose of structuring and making meaningful decisions.

It is important to ensure that today's adult students have both the broad scientific background necessary for understanding the problems of the future and the ability to address difficult decisions in a systematic way. It is time to expand curricula on numeracy to include decision making and, simultaneously, to expand decision-making curricula beyond the university walls to make the ideas accessible to all. By teaching the basics of sound decision making as part of every adult student's curriculum, we can provide guidance by teaching people how to think rather than just what to think. We also can help to establish a direct link between personal values and the creation of satisfying options, whether in an individual or social setting. With the help of improved training in decision making, our nation's citizens will gain a better grasp of their capabilities and be in a better position to make the difficult scientific and social choices that lie ahead.

REFERENCES

Baron, J., & Brown, R. (1991). *Teaching decision making to adolescents.* Hillsdale, NJ: Erlbaum.

Beyer, B. K. (1985, April). Critical thinking: What is it? *Social Education,* pp. 270-276.

Beyth-Marom, R., Dekel, S., Gombo, R., & Shaked, M. (1985). *An elementary approach to thinking under uncertainty* (S. Lichtenstein, B. Marom, & R. Beyth-Marom, trans.). Hillsdale, NJ: Erlbaum.

Clemen, R. T. (1996). *Making hard decisions: An introduction to decision analysis.* Belmont, CA: Duxbury.

Ennis, R. H. (1962). A concept of critical thinking. *Harvard Educational Review, 32,* 81-111.

Gregory, R. (1991). Critical thinking for environmental health risk education. *Health Education Quarterly, 18,* 273-284.

Gregory, R., Clemen, R., Satterfield, T., & Stone, T. (1996). *Creative decision making: A curriculum materials guide for secondary-school educators.* Available from Decision Research, 1201 Oak Street, Eugene, OR 97401.

Keeney, R. (1982). Decision analysis: An overview. *Operations Research, 30,* 803-839.

Newmann, F. M. (1990). Higher order thinking in teaching social studies: A rationale for the assessment of classroom thoughtfulness. *Journal of Curriculum Studies, 22,* 41-57.

Plous, S. (1993). *The psychology of judgments and decision making.* New York: McGraw Hill.

II

APPROACHES TO INSTRUCTION

5

Instructional Strategies for Adult Numeracy Education*

Lynda Ginsburg
University of Pennsylvania

Iddo Gal
University of Haifa

Many adult educators want to improve the ways in which they teach mathematics to adult students, particularly in light of recent reforms in mathematics education at the (K-12) school level, as well as what we know about how adults learn and use math. This chapter presents 13 principles that have the potential to improve adult numeracy education and discusses specific instructional practices and strategies to guide implementation of these principles.

---

*This work was supported by funding from the National Center on Adult Literacy (NCAL) at the University of Pennsylvania, as part of the Education Research and Development Center Program (Grant No. R117Q0003) as administered by the Office of Educational Research and Improvement, U.S. Department of Education. The opinions expressed here do not necessarily reflect the position or policies of NCAL or the other supporting agencies.

INTRODUCTION

During the past 20 years, the field of mathematics education has been enriched by extensive research into the nature of the cognitive processes involved as people learn mathematics (e.g., see Siegler, 1991; Wearne & Hiebert, 1988) and teaching techniques that are particularly effective in helping students learn mathematics (Ball, 1993; Lampert, 1986). The National Council of Teachers of Mathematics has drawn implications from this research, as well as from teachers' experiences, and produced the Curriculum and Evaluation Standards for School Mathematics (1989), a vision of math education that is revitalizing and revolutionizing the teaching of math in the United States (see chap. 3).

Although the focus of the *Standards* is on the K–12 curriculum, adult numeracy educators can certainly identify with and learn from NCTM's articulation of the broad goals of math education and the descriptions of instructional practices aimed at achieving those goals. However, the goals and instructional practices of adult numeracy education may not always be the same as those in the K–12 realm (Gal, 1993). Further, the life situations, prior educational and work experiences, and motivations of adult students are very different from those of the children and teens in traditional schooling environments. These factors shape the need for or perceptions of the usefulness of new learning that adults bring to their studies; they may thus have an important impact on adults' willingness to come to, invest in, or stay in educational programs.

Adults return to educational settings when they feel they will benefit from that education. It will help them perform tasks or deal with real-world challenges that they are presently not ready to handle. Furthermore, they come with a wealth of experiences that are closely tied to their self-identity. Although these experiences may differ among students and contribute to heterogeneous classes (see chap. 6), they are also a rich resource for enhancing instructional episodes (Knowles, 1990). On the other hand, years of experiences may also have created deep-seated negative beliefs and attitudes (toward learning in general and mathematics learning in particular) and ingrained strategies and ideas that may be counter-productive or actually wrong (Ginsburg, Gal, & Schuh, 1995).

Theories of adults' learning processes (e.g., Knowles, 1990) tend to be general in nature and do not address learning in specific content domains. This chapter suggests specific instructional practices and strategies that are aimed at improving numeracy education for adults. The suggestions reflect what we have learned about how adults learn, how children and adults learn math, and what is important math learning (important both in the sense of building a foundation on which further math learning can be constructed and in the sense of real-world utility to individuals). Chapters 1 and 2 provide a more detailed discussion of some of these backgrounds. Some of the suggestions made here are also discussed and elaborated on elsewhere in this volume; references to related chapters are included.

The suggestions in this chapter are organized as 13 interrelated principles, summarized in Figure 5.1. These principles aim to help students develop particular numeracy practices and skills that are important and useful to adults as they function in the world. These principles assume acceptance of a broad conception of what numeracy education encompasses, as described in Chapter 1. They aim to contribute to meaningful and useful numeracy learning, while keeping in mind that the development of functional skills, conceptual understandings, and positive dispositions is important to support further learning and effective application of mathematical knowledge.

The community of adults learning math is a diverse one, as Chapter 6 describes, and there is a multiplicity of educational contexts in which adults study. We thus present instructional principles and suggestions in broad terms, expecting and encouraging practitioners to select, modify, and apply these ideas to their own local situations.

1. Address and evaluate attitudes and beliefs regarding both learning math and using math.

2. Determine what students already know about a topic before instruction.

3. Develop understanding by providing opportunities to explore mathematical ideas with concrete or visual representations and hands-on activities.

4. Encourage the development and practice of estimation skills.

5. Emphasize the use of "mental math" as a legitimate alternative computational strategy and encourage development of mental math skill by making connections between different mathematical procedures and concepts.

6. View computation as a tool for problem solving, not an end in itself.

7. Encourage use of multiple solution strategies.

8. Develop students' calculator skills and foster familiarity with computer technology.

9. Provide opportunities for group work.

10. Link numeracy and literacy instruction by providing opportunities for students to communicate about mathematical issues.

11. Situate problem-solving tasks within meaningful, realistic contexts in order to facilitate transfer of learning.

12. Develop students' skills in interpreting numerical or graphical information appearing within documents and text.

13. Assess a broad range of skills, reasoning processes, and dispositions, using a range of methods.

Figure 5.1. Instructional principles in adult numeracy education

Address and Evaluate Attitudes and Beliefs Regarding Both Learning Math and Using Math

Rationale. Many students come to the classroom with fears about their own abilities in the area of math ("I can't do math," "I can't remember how to do math even though I've learned it so many times," etc.). In addition, students often carry with them nonproductive beliefs about what it means to "know math" or what learning math should look like, such as "there is always only one correct answer," "there is one right way to solve problems," "you should always work alone on math problems," and so forth (McLeod, 1992). These negative attitudes and beliefs often hold students back from engaging in math tasks in meaningful ways and from trusting their own mathematical intuitions. Limiting self-images and beliefs are particularly harmful because students may inadvertently communicate them to their own children.

For many, negative attitudes toward mathematics initially develop in response to early school experiences of failing to understand. A student's initial confusion may be followed by a failure to receive further explanations or assistance from the teacher, leading to a loss of confidence and panic over the loss of control of his or her own learning. As these experiences are repeated, the student feels frustrated and assumes it is futile to expect understanding. Finally, the student "switches off" to distance him- or herself from the situation (Allan & Lord, 1991; Tobias, 1994).

Suggestions. It is important to discuss openly with students how traditional K-12 methods of teaching math and stated or implied messages from parents and teachers may have caused them to develop negative beliefs and attitudes; otherwise, some students may continue to blame themselves and not approach classroom-based and real-world mathematical tasks in a productive way. Have students freely talk or write in a trusting environment about their attitudes and beliefs. The teacher can share his or her own fears and experiences. Point out and encourage students to look for manifestations of their existing (even if informal) mathematical understandings, of which they may be unaware, to encourage the development of feelings of comfort and control. This process of exploration and reflection should occur throughout instruction, not only at the outset, as different negative attitudes and beliefs may be tied to different areas of the curriculum.

Determine What Students Already Know About a Topic Before Instruction

Rationale. Adult students have a rich background of real-life experiences. Even if they have learned little formal math, they are likely to have engaged in counting, sorting, measuring, playing games of chance, and, most importantly, handling money (Bishop, 1991). Through such experiences, adults most likely have developed various skills and their own (partially) formulated conceptual understandings, some correct and some incorrect. For example, Ginsburg et al. (1995) investigated adult literacy students' understandings of five basic concepts having to do with percent (such as 100% means "whole" or "all," percentages of the parts of a "whole" add up to 100%, percents lie on an ordinal scale, etc.); almost all students had some reasonable ideas about some of the concepts, but these were accompanied by gaps in understanding, with different students displaying different gaps.

Identifying students' partial understandings and intuitions is important because new learning will be filtered through and become integrated with prior knowledge. Each student's informal knowledge should be identified so that new instruction can be designed to link with what already has meaning to the student. At the same time, attention must also be paid to incorrect ideas or "patchy" knowledge so that these do not distort new learning or cause confusion.

Suggestions. An informal discussion of students' real-world and school experiences should be used to explore "What do we know already?" when starting a new area of instruction. This gives students opportunities to think about and discuss mathematical issues and to begin linking bits of disconnected prior knowledge into a structure without the competing demands of computation. A computation pretest or a "test of sums" (either a standardized commercial test or one constructed by a teacher) should not be used as a primary source of information. Besides producing anxiety for the student, it reinforces a sense that mathematical tasks involve only computation. In addition, such tests rarely provide teachers with much information on students' underlying thinking processes and informal knowledge. The use of such tests, if any, should be supplemented or replaced by an examination of students' reasoning strategies and informal number skills, in formal but, more importantly, in realistic, familiar contexts. (See chap. 16 for an illustration of a methodology for assessing informal strategies, as well as further discussion and suggestions regarding assessment practices on pages 108-109 in this chapter.)

Develop Understanding by Providing Opportunities to Explore Mathematical Ideas with Concrete or Visual Representations and Hands-on Activities.

Rationale. Being able to create or manipulate a physical model enables students to visualize the concrete reality underlying abstract symbols and processes. Students will find the math they are learning more meaningful if they can link the ideas, procedures, and concepts to realistic situations, concrete representations, or visual displays; these can help students "see" and "feel" how and why computational algorithms work. In addition, the use of concrete materials or visual displays may provide students with a means for monitoring their own computations and procedures, helping them to pinpoint gaps in understanding when they might have difficulty identifying and describing these gaps using more abstract language. The process of working with hands-on activities and with models may help students develop backup strategies that can be used when they become confused with the mechanisms of newly learned strategies or when they want to be certain that computations are indeed correct. However, Ball (1993) warns that although the particular concrete objects or representations the teacher selects should highlight important mathematical content, the teacher should also be aware that there may be some limitations within the model. For example, "using bundling sticks to explore multidigit addition and subtraction directs attention to the centrality of *grouping* in place value, but may hide the importance of the positional nature of our decimal number system" (p. 163; emphasis in original).

Suggestions. During the learning process, students should move between "real objects" (e.g., apples, coins, cups), "representative objects" or "representations" that can stand for real objects (e.g., blocks, toothpicks, and beans, or drawings and diagrams of real objects), and abstract symbols (e.g., numbers). At the beginning of a new unit, have students solve a number of related problems using real objects or a representative model of a situation. Encourage students to talk about these examples, make observations, and explain how mathematical processes (e.g., counting, adding, taking doubles or halving) work, before introducing formal notations or mathematical rules. After students are comfortable with new concepts in a concrete context, a discussion generalizing and formulating more abstract principles is appropriate. (See chap. 7 for illustrations of these ideas in arithmetic with numbers up to 100.)

Many commercial publishers sell "manipulatives" such as pattern blocks, base ten blocks, fraction circles, Cuisenaire rods, geoboards, and so on, for use as representative objects in teaching specific topics.

Transparent versions of these objects are also available for use with an overhead projector and can greatly help in demonstrating work with manipulatives to groups of students. These materials are handy to use because the sizes, shapes, or color schemes are consistent and "user friendly." In addition, everyday objects such as toothpicks, beans, cards, nails, buttons, cardboard pieces, and so forth can also serve as inexpensive manipulatives.

Encourage the Development and Practice of Estimation Skills

(Note: This section and the one following discuss estimation and mental math, two topics that are related in some aspects, support each other, and are sometimes used for the same purposes. Here they are being discussed separately because they also may be used in different situations, have different pedagogical justifications, and are for the most part different procedures.)

Rationale. Many everyday or work tasks do not require precise, computed answers, but rather quick approximations of numbers, distances, time-frames, and so on, based on some known information. The demands of the situation will determine how precise or imprecise an estimate has to be and whether it can be made mentally or requires written work or a calculator. When shopping, it is often more reasonable to approximate a total cost mentally instead of using a calculator or doing a written computation to determine if one is carrying enough cash to pay for groceries, or to confirm that the total at the cash register is correct. Computing a 15% tip at a restaurant does not have to be precise, yet it can be an overwhelming task to someone who can only conceive of calculating percentages by using a formula. Generating quick, approximate answers to math questions on standardized tests is often sufficient to discount all but one or two of the response choices. Good estimation skills can also be used to catch gross computational errors from misplaced decimal points or from errors made while using a calculator (Schoen & Zweng, 1986).

Suggestions. Students should be asked to identify everyday and work situations for which estimates may be more appropriate than exact answers, to reinforce the notion that estimating is a valuable skill, not merely something one does when one does not know how to compute an answer the "right" way. This may also be used to address the (inappropriate) belief held by many students that there should be only one "correct" way to respond to a problem involving mathematics. Discuss reasons for the need to estimate and the inevitable tradeoff between the benefits of estimating (e.g., time saved) and the possible cost in error due to lack of precision. Stress that there are no "right" or "wrong" estimates, only ones that are closer or farther from a computed answer and that the importance of the degree of exactness (which may determine the estimation strategy to use) depends on the requirements of the situation. Note that Sowder and Wheeler (1989) found that there is a developmental progression in accepting that there can be multiple good estimates and that computational estimation requires "rounding-then-computing" rather than "computing-then-rounding."

Students can share with each other the estimation strategies they use (such as "multiplying by 10 instead of 9 and then subtracting a little"), and the teacher can supplement the class repertoire with strategies he or she uses in everyday life. Differentiate between rounding numbers as a strategy for estimating and "rounding off" an answer that has been computed. Encourage students to adopt a reflective attitude toward estimation, asking themselves questions such as (a) How reasonable (close) is the estimate? (b) Is this degree of accuracy appropriate for purposes of the situation? (c) Was the method or strategy appropriate? (d) Does the situation suggest making a high or low estimate (e.g., when making sure I have enough money for groceries, I may want my estimate to err on the high side)?

For classroom estimation practice, students can be asked to estimate amounts as they would in the everyday contexts for which an exact answer is not normally required (i.e., tips, number of miles per tank of gas, approximate cost of a basket of groceries, number of miles between New York and Chicago as represented on a map, and time it will take to drive from one to the other). They can also estimate answers to traditional arithmetic exercises before carrying out the computations (on paper or with a calculator) as a way to check reasonableness of formal computations. Using sample standardized test items with multiple choice responses for estimation practice gives students an opportunity to see that often a good estimate can replace tedious computation and point to a reasonable response.

Emphasize the Use of "Mental Math" as a Legitimate Alternative Computational Strategy and Encourage Development of Mental Math Skill by Making Connections Between Different Mathematical Procedures and Concepts

Rationale. Mental math involves conceiving of and doing mathematical tasks without pencil and paper (or a calculator). Its key element is using "in my head" mathematical procedures that may be quite different from school-based, written procedures. In everyday life, we often want to know precise answers, not estimates, to questions involving numbers, but we do not want to stop to write a computation, or we realize that the written procedure is cumbersome, while a solution is easily attainable in another way. (Ask your students and colleagues *how* they multiply $3.99 times 4 [e.g., when buying 4 items at the supermarket] in their heads, and, after they respond, inquire why they do it differently than on paper.) Mental math is not simply carrying out the mental equivalent of paper-and-pencil algorithms, but rather flexibly using diverse strategies determined by the properties of particular numbers or the relationships between quantities within given problem situations (Hope, 1986).

There are two reasons for encouraging the development of mental math skill in the classroom: its practical usefulness and its educational benefits. Often a quick response to an everyday or workplace number-laden situation is expected or required. Although computation using standard algorithms is efficient if paper and pencil is used, applying those standard algorithms to mental computation is awkward at best and at worst creates confusion and a memory overload (e.g., when having to conduct long division, or multiplication of two digit numbers). Our short-term memory has limited capacity, and keeping numerous intermediate results of calculations simultaneously in mind is very difficult. Mental math strategies aim to reduce the mental load and the amount of information necessary to keep or process in mind at one time, thus reducing errors and improving accuracy.

Perhaps more important is the educational benefit from learning to do mental math, as it helps develop a facility for moving between equivalent representations of quantities. That facility requires knowledge of how and why procedures work, and an expectation that there are meaningful connections between concepts and an understanding of such connections. Traditionally, teachers have assumed that students will intuitively make connections between the different topics encompassed in math classes, but, in fact, students often think of them as a series of discrete and unrelated topics. (This perception is reinforced by individual "topic" workbooks like those for "fractions" or "percents"). Students who expect there to be connections between different mathematical concepts will be less fearful of learning math because they will expect new mathematical ideas to be extensions of what they already understand.

Adult students, in particular, may benefit from developing mental math skills as alternatives to the computational algorithms that are typically taught because they may have difficulty remembering multistep procedures that are not really meaningful to them. Standard computational algorithms do not correspond to the ways people naturally think about numbers and therefore often require suspended understanding (Plunkett, 1979). In contrast, mental math procedures are flexible and can and should be adapted to suit the particular numbers involved; they require a person to select, adapt, or invent an appropriate procedure and work with "easy" numbers or selected amounts in ways that make sense to the actor. Mental math strategies often yield an early approximation of the answer to a problem because usually the leftmost digits are used first in the calculations (contrary to how addition, subtraction, and multiplication work first on the rightmost digits). Thus, the extent to which students can do mental math can also show their understanding of and "sense" of the number system and shed additional light on their mathematical skills.

Suggestions. Some students are fearful of or uncomfortable with doing mental math ("This is not the real math I learned in school") and may not spontaneously generate or want to use mental math strategies. Therefore, mental math strategies should be identified, discussed, and then practiced over time in numerous situations so that students will trust their abilities enough to use them when appropriate. Because some numbers are less amenable to mental math than others (the exact answer to $3.99 times 4 can easily be determined using mental math, but multiplying $3.62 times 4 is more awkward), students should become aware of and monitor their own decisions regarding strategy use. It is important that students are able to discuss and demonstrate their understanding of the differences and similarities between what they figure out in their heads and what they do with paper and pencil and that they develop a repertoire of mental math strategies with which they feel comfortable.

Refer frequently to previously studied material to help students see the connections between different mathematical concepts, such as fractions, decimals, and percents, so that students become flexible switching from one "system" to another when performing "mental math" (e.g., "25% off" means that one fourth of a price should be subtracted). Describe mental math strategies you use yourself and spend time practicing mental math skills first with small numbers and then with larger or more complex numbers. Discuss *why* you can get the same answer using different computational procedures or representations and elicit opinions about the relative advantages of one representation or mental or written procedure over another.

View the Mastery of Procedures as a Tool for Solving Problems, Not an End in Itself

Rationale. Even though the acquisition of procedural skills (e.g., doing two-digit multiplication, subtraction, division, converting percents to decimals, graphing) is important, it is of little use unless students also develop the ability to determine *when* certain procedures are appropriate or useful, and *why*. Time spent mastering procedural skills should be balanced with time spent talking about the uses of such procedures and enabling students to grapple with the application of their skills in both familiar and less familiar situations. Learning of procedural skills has to be situated in multiple and reasonable contexts, so that the skills mastered become generalizable (Strasser, Barr, Evans, & Wolf, 1991), and to increase the chance that skills are actually used in useful ways outside the classroom.

One approach to problem solving in which the complementary role of procedural skills, especially computations, becomes apparent is suggested by Brown and Walter (1990). These authors argue that it is important to explore the reasons a question is asked (or a problem posed) in the first place and then to explore ways in which small changes in the problem (asking "What if . . . ?") might affect solution approaches and the kinds of procedural skills applied. Through this approach to problem solving, computations (or geometric work, graphing, or other procedures) are done to answer questions, not for their own sake, and their utility (compared to estimates or other procedures) can be examined. By extending problem-solving experiences through "what if" questions and simulations, critical thinking skills and deeper understanding can be developed. Clearly, procedures are but a tool for answering "important questions," just as word decoding is a tool in the process of finding meaning in written communication.

In addition, students need to develop a sense of *why* a particular computational procedure is appropriate in a particular situation. This requires an understanding of what is happening as a computational procedure is being used and of the differences and commonalities between procedures. Two simple examples are the connections between multiplication and addition or between division and subtraction. The understanding described here is not equivalent to a recitation of the steps involved in performing a computational procedure, but rather focuses on the student's internalized vision of a model (which can be described in concrete or representational terms) underlying a procedure.

Suggestions. When teachers (and textbooks) focus on computational procedures in isolation from meaningful contexts, students assume they "know" the mathematical content and that this "knowledge" will be useful to them. However, seldom are decontextualized computational skills useful (perhaps only on standardized tests). Rather, most real-world situations require application of procedures and computations only within the context of problem solving. Take, for example, the case of fractions. Often, at least in the United States, students "learn" how to add, subtract, multiply, and divide fractions, completing numerous worksheets with context-free numbers separated by designated operations. Once the procedural skills are mastered, fractions have been "learned." Yet, when students have to determine the quantity of eight-foot boards that must be ordered so that five bookcase shelves of two and three-fourths feet can be cut, they ask, "Do we add, subtract, multiply, or divide?"

By focusing a significant part of instructional time on situational questions, students will have opportunities to analyze problem situations and appreciate the function of computational procedures (or the use of estimates, or mental math) as part of the problem-solving process. A variety of computational procedures can be practiced when problems are extended or placed in realistic contexts. The bookcase problem can be extended by asking "What if?" questions such as, "What if we have 2 eight-foot boards and want to make as many 3-foot shelves as possible?," or, "What bookcase might we create if we want to use all the wood from 2 eight-foot boards for five shelves that do not have to be the same size and will be supported by cinderblocks?" This approach has an additional benefit of providing contexts for discussions of alternative solutions.

Another way to reinforce connections between computational skill and applications is by asking students to write their own problem stories or word problems (which is the focus of chap. 10), targeting a particular procedure (i.e., multiplying fractions). Student-generated problems can be shared with other class members, mixing problems suggesting different computational procedures so that students will have opportunities to select appropriate solution methods rather than think, "This week we have been learning to multiply fractions, so you must have to multiply to solve each of these fraction problems." The opportunity to use manipulatives (realistic and representational) and other visual aids while solving problems will help students develop a sense of why one procedure works, whereas another is inappropriate or less efficient.

Encourage Use of Multiple Solution Strategies

Rationale. Some students come to instruction believing that there is only one proper or best way to solve a math problem. When students feel that they do not have the specific knowledge or skill needed for a particular problem, they may become anxious or frustrated and quickly give up rather than persevere and look for alternative paths to solutions. Students will become flexible problem solvers by developing webs of interconnected ideas and understandings; instruction should include explanations and demonstrations of many ways to arrive at a good solution to a problem and continuously point to connections between mathematical representations, concepts, and procedures (Hiebert & Carpenter, 1992). In addition, different strategies or representations may be more meaningful to some students than others. By having alternatives available, more students will be able to connect new learning with their individual experiences and perspectives.

Suggestions. Rather than leaving a problem as soon as it has been solved and rapidly moving to the next one, ask if any student can think of another way to approach it. Students need time to question and discuss the meaning, accuracy, or efficiency of answers or methodologies. Frequently ask students why they did what they did and what they could have done as an alternative. Include mental math strategies when enumerating alternative solution paths. Continually cycle back to previous topics to show connections between new and old skills and concepts.

Develop computational algorithms logically so students see that the algorithms are simply shortcuts for time-consuming procedures (such as multiplication for repeated addition and division for repeated subtraction) or alternatives for other representations (as percents for fractions). While exposing them to new methods, allow students to continue to use "lower level" strategies (such as finger counting when adding or subtracting and multiple addition rather than multiplication) if they need to do so; most will eventually see that certain strategies are cumbersome and, as they feel more secure with new strategies, will probably begin using the "higher level" strategies. If they never get to that point, they will at least be able to use a strategy that is dependable and meaningful to them.

Develop Students' Calculator Skills and Foster Familiarity with Computer Technology

Rationale. These days, there is less need for skill in fast calculation given the availability of calculators. This is not to say that calculators should replace computation skills, but rather that the goals of adult numeracy education will be enhanced by encouraging judicious calculator usage. Students should have opportunities to become skilled at using what has become an accepted and essential workplace tool. In addition, calculators can be used as an instructional tool; students can quickly observe the results of many calculations, see patterns, make generalizations about mathematical processes, and focus on understanding without getting bogged down in lengthy calculations (Manly, 1997-98).

Some teachers fear that allowing students to use calculators will reduce their opportunities to perfect traditional computational skills. Although calculators may take the place of tedious computation and ensure accuracy, they do not replace deep understanding of mathematical concepts and procedures, and they cannot make decisions or solve problems. It is becoming ever more important to know *what* a procedure does, *why* it works, and *how* the results can be evaluated to make certain they are appropriate responses to the original task. Such demands are essential whether students learn to compute with or without calculators.

A review of 79 research studies of students ranging from kindergarten through twelfth grade showed that, in almost all grades, the "use of calculators in concert with traditional mathematics instruction apparently improves the average student's basic skills with paper and pencil, both in working exercises and in problem solving" (Hembree & Dessart, 1986, p. 86). In addition, for students in all grades and at all ability levels, attitudes toward mathematics and self-concept in mathematics were better for those who used calculators than for those who did not.

Beyond issues related to calculators, adult educators must consider the broader issue of the use of computers. As most jobs now require some familiarity and facility with technology, many students are eager to use computers. Spreadsheets and simulation software provide engaging environments that support development of number sense, problem-solving, and critical thinking skills (Jonassen, 1996). Aside from having opportunities to encounter mathematical content with these tools, students develop computer literacy and computer-related skills that are valued by the community and the workplace. (Chap. 8 presents an extensive discussion of technology and numeracy education.)

Suggestions. Provide opportunities to use calculators and set aside time to make certain that each student knows how to use his or her calculator, being wary that sometimes functions work differently on different calculators. The calculators can then be used for various types of activities, such as (a) checking mental or written calculations; (b) observing patterns resulting from numerous repetitive computations that would be tedious if done manually (e.g., investigating what happens when two decimal fractions less than 1 are multiplied together or what 10% of any number is), or for processing results from experiments and drawing conclusions from patterns; or (c) problem solving if and when a student feels it would be a helpful aid. By posing realistic, extended problems, as suggested earlier, students will have opportunities to determine for themselves when use of a calculator (rather than estimating, using mental math strategies, or written computation) is the most appropriate thing to do. The comparative benefits of these options for different problem situations can also be discussed in class.

Explore the possibility of using computer software to help students develop specific mathematical skills, keeping in mind that computer usage should be integrated with other classroom activities and accompanied by classroom discussions. As described in Chapter 8, there are different ways to integrate the use of computers in numeracy education. Students could use, for example, simple "integrated" software (with graphing, database, and spreadsheet capabilities) as an aid in planning, managing, and presenting results of group projects, or simple word processing programs to write math journals and explanations of solutions to extended problems. Such practices can enrich students' experiences with literacy-numeracy connections and help them integrate their emerging skills; printouts should find their way into students' portfolios.

Provide Opportunities for Group Work

Rationale. The SCANS Commission (1991; and see chap. 3) suggests that those joining the workforce must be competent in working with others on teams, teaching others, and negotiating. Often the contexts in which these skills are demanded include problem solving or communications involving numerical information. Traditionally, math has been studied by oneself, and communal work was relegated to other disciplines such as science or social studies. Yet, in the real world, people regularly have to communicate about numerical issues (negotiating a contract, making business or purchasing decisions, defending an estimate, etc.). Furthermore, students often benefit from their peers' observations or explanations because one student may be able to identify another student's point of confusion or explain a concept with examples that are especially helpful for that particular student (Slavin, 1990). Therefore, cooperative work can contribute to development of cognitive skills (Dees, 1991).

Suggestions. It is difficult to expect students to develop realistic group work skills if instruction revolves around traditional, isolated, brief tasks of the kind espoused in most textbooks. There is a need to create an atmosphere in which students frequently have to work together and help or teach each other. Periodically develop long-term projects revolving around realistic and engaging problems for which heterogeneous and extended group efforts are appropriate, such as organizing a group trip, arranging a party or a meal (including planning, deciding on and managing schedules, budget, supplies, materials, division of labor, etc.), or conducting a survey about a meaningful issue (including collecting, analyzing, and reporting on findings and implications).

Groupings may vary according to the particular tasks involved and consist of pairs of students, small groups of three or four students, or even a large group with subgroups, each of which would address a particular aspect of a complex task and report findings to the larger group to inform a decision-making process. Ground rules for behavior within groups should be established as a class, with students involved in determining appropriate interactions.

Link Numeracy and Literacy Instruction by Providing Opportunities for Students to Communicate about Mathematical Issues

Rationale: Many workplace and real-world situations require individuals not only to solve mathematical problems but to communicate their reasoning and the results or implications of their work to others (Carnevale, Gainer, & Meltzer, 1990). Adults also frequently find themselves discussing mathematical concepts with their school-aged children as they help with homework assignments or study for tests. Communicating mathematically might include drawing a diagram (of a room to plan carpeting), writing a letter about an error on a utility bill, calling someone to report that a shipment arrived with less than the ordered amount, or negotiating terms of a sale, and so forth. Mathematical communication may also involve "passive" listening and critical comprehension processes, as when adults have to make sense of statistical information embedded in a newspaper article (e.g., about results from a recent poll), or when making sense of TV reports regarding medical, environmental, or crime-related news (Gal, 1999).

Thus, reading, writing, and communicating are activities within which math is found, and they should be taught and practiced with mathematical content. "Talking about math," whether verbally, schematically, or in written form, enables students to clarify and structure their thinking so that a target audience will clearly understand their information or argument. A focus on communication issues in the context of learning mathematics contributes to and should be seen as an integral part of developing statistical, scientific, and general literacy skills.

Suggestions. In the course of mathematics instruction and problem solving, encourage students to put into words for others what they are doing and why, or to discuss their thoughts about the meaning of other people's work or arguments, using both written and oral formats. As discussed in detail in Chapter 10, there are multiple ways to enhance mathematical communication and literacy skills when learning mathematics. These include, for example, the writing of project journals and reports, learning logs, math stories, letters to the editor, or preparation of a newsletter, which overall encompass a range of literacy experiences and literary products.

Written and verbal communication skills can also be developed through classroom activities that do not dictate a single solution process or lead to a single right-or-wrong answer, but rather require students to form, present, and justify opinions. These may involve a range of issues that can be tailored to the particular skill levels, needs, and interests of students, and that in general fall into four categories:

- math-based problems, such as "How can we measure the area of an irregularly-shaped lake?
- statistical projects, such as a survey (involving generation of data and presentation of analysis and conclusions)
- interpretive tasks involving data-based or statistical arguments, such as "What is your opinion about the argument in this newspaper article, that the establishment of the death penalty in some states has contributed to a reduced rate of serious crimes?"
- decision projects that require the presentation of recommendations for a course of action, such as suggesting a plan for a class trip, or proposing what computer system to buy, what apartment to rent, or what insurance to purchase. Such projects involve background research about competing courses of action and discussion of the pros and cons of options, in terms of cost, timelines, resources needed, and so forth.

Situate Problem-Solving Tasks Within Meaningful, Realistic Contexts in Order to Facilitate Transfer of Learning

Rationale. Educators hope that the skills that students develop in the classroom will be used effectively and appropriately in out-of-school environments. Unfortunately, researchers have found that skills learned in one environment, may not be easily transferred to or applied effectively within another environment and the farther the learning context is from the target context, the less likely it is that transfer will occur (Mikulecky, Albers & Peers, 1994; Nunes, Schliemann, & Carraher, 1993; Perry, 1991). For example, Ginsburg et al. (1995) found that some adult students who were able to compute solutions to context-free percent exercises (such as "25% of 20=") by applying a standard school-based multiplication procedure were unable to determine how much money they would save if an $80 coat were on sale at "25% off."

It is thus important to situate problem-solving tasks within meaningful, realistic contexts in order to facilitate transfer of learning.

Students should routinely be given opportunities to practice using their new skills in environments that are similar to the life and work environments in which they will have to function, rather than just in context-free environments such as those found in workbooks with extensive isolated arithmetic practice exercises. In addition, interest in learning will be sustained if the students can see clearly that what they are learning will be directly applicable to situations in their own lives, and they will feel more empowered to use such skills.

Suggestions. Elicit students' experiences in situations in which mathematical issues arise and use them to develop meaningful, realistic contexts for problem-solving tasks. For example, if students live in an area that has some availability of public transportation, a project could be developed around whether to buy a car. Considerations of cost, car payments including interest, upkeep, insurance variables, and frequency of usage can be compared with public transportation costs and limitations. Students, as well as teachers, can be involved in posing questions and asking "What if?" to explore consequences of alternative scenarios. Simulations based on students' experiences can help them practice applications in contexts that are different (in planned ways) from textbook or "school-like" situations. Encourage students to reflect on what is different between school-like and real-life problem solving, in terms of skills, degree of accuracy expected, tools used, assumptions about the situation, or world knowledge and beliefs they bring into the process.

In designing realistic tasks, teachers should make sure to attend to the difference between real-world problems and made-up problems that only use real-world data. For instance, computing and comparing the price per pound/kilo of a product available in both small and large packages (such as cereal, paint, etc.) represents a real-world problem that is appropriate and useful for adults. Finding the average price of 6 items in a grocery cart is not a meaningful problem for adults. The first task (product comparison) simulates what one would actually want to know in a particular situation, whereas the second task (averaging) uses a real-world context as an excuse for a mathematical task that is not meaningful within this context. Unfortunately, problems of this second type often come up in many textbooks and may mislead teachers and students to assume they are developing effective functional skills.

Develop Students' Skills in Interpreting Numerical or Graphical Information Appearing Within Documents and Text

Rationale. Numerical information is often embedded in text-rich contexts that are of importance in adults' lives, such as statements of employee benefits, payment schedules, contracts, tax instructions, or maintenance agreements (Kirsch, Jungeblut, Jenkins, & Kolstad, 1993). People often read newspapers or magazines and have to interpret graphs or statistical information presented in tables or text. These tasks are unique in that most often there is little or no computation to do, but only a need to apply a conceptual understanding of diverse mathematical topics and combine this understanding with text comprehension. If adults skip pertinent numerical information because they feel uncomfortable or incompetent about processing it, the text loses meaning and people lose access to critical information. The goals of both numeracy and literacy instruction can be enhanced when students are helped to develop statistical literacy and interpretive skills (Gal, 1993).

Suggestions. Have students graph information from their lives, such as a circle graph showing how they spent the last 24 hours, or a bar graph showing how many cigarettes they smoked or cups of coffee they drank each day for a week. Students will see the connections between events in their lives and the graphs they create; these experiences help students understand how other graphs are constructed and the information that can be gleaned from them.

Students could bring in and read newspaper articles or other text-rich materials containing numerical information that must be interpreted but not necessarily "computed" and report a summary of that information orally or in writing to other students. Alternatively, all students in one class could read the same article or document and discuss its implications. In these discussions, it is necessary to be certain that students understand the vocabulary and comprehend any technical terms.

Model and encourage the development of a critical stance when making sense of articles or advertisements that draw conclusions from summarized data (e.g., "Nine out of ten doctors prefer..."). During class discussions, push students to challenge presented information and implications by asking probing questions such as:

- Where did the data on which this statement is based come from? How reliable or accurate are these data?
- Could the study have been biased in some way?
- Are the claims made here sensible and justified from the data?
- Is there some missing information?

Students may initially feel too intimidated to take a challenging attitude to what is printed or reported in the media, but they will eventually begin to feel comfortable in the role if they have opportunities to practice asking these critical questions.

Assess a Broad Range of Skills, Reasoning Processes, and Dispositions, Using a Range of Methods

Rationale. Educators communicate their pedagogical priorities to students in part through the assessments they use. Many adult education programs, at least in the United States, use tests that employ multiple-choice formats to evaluate the mathematical skills of incoming students or to assess learning gains. By using *only* such tests, it is communicated that what is valued in numeracy education is mostly the ability to compute with decontextualized numbers or to solve brief (and sometimes contrived) word problems. Yet, if we accept the curricular goals and instructional principles discussed earlier and in other chapters of this volume, we should significantly extend the scope and methods of assessments used in adult numeracy education.

Assessments should focus on worthwhile content that reflects the instructional goals of the students' program of studies. Problems and tasks used in assessments should yield information that can provide meaningful feedback to the student, as well as inform instructional decisions by the teacher. In addition to mastery of computations and formal procedures, assessments should diagnose the many additional skill and knowledge areas that are part of "being numerate," such as students' ability to interpret statistical and quantitative claims, act on numerical information in technical documents and forms, apply mathematical reasoning and explain it, solve realistic problems, communicate about mathematical issues, and so forth. (See Lesh & Lamon, 1992; MSEB, 1993; NCTM, 1995; Webb, 1992, for general discussions of diverse assessment issues in mathematics education, as well as chaps. 15 and 16, this volume, for a focus on assessment issues in adult numeracy education.)

Suggestions. The upshot of the ideas, assumptions, and suggestions outlined earlier in this chapter is that numeracy assessments should extend well beyond examining students' ability to find the right answer for a computational exercise. Assessments should also include open-ended, extended tasks that may have more than one reasonable solution and/or solution path and require that students *explain* their reasoning and the significance of their solutions. Assessment tasks could be brief or extended and culminate in diverse products that may combine graphs, tables, drawings, written text or arguments, results of computations, or oral reports. To arrive at or communicate their findings and thoughts, students should be able to use paper and pencil, computer outputs or images, calculators, manipulatives, and models. Assessment products could describe a solution process (of an individual or a group), present recommendations for a course of action, be in the form of simulated memos or letters that aim to communicate about mathematical issues with specific real-world audiences (and demonstrating both appropriate mathematical know-how and literacy skills), or be comprised of performances on simulations of realistic tasks relevant to a particular student population.

This vision of a range of assessment tasks, support tools, and outcomes recognizes that it is important to have information about both the process and the product of students' work in order to be able to evaluate their understanding of and ability to apply mathematical concepts. By expanding the notion of what constitutes "assessment," educators can integrate assessment and teaching and employ teaching activities to generate information that can satisfy both diagnostic and instructional needs. To the extent that teachers and students together find such information of value, representative samples of student work can be recorded and stored (e.g., in portfolios or other forms) for various uses. However, unless assessment information is put to good and timely use by teachers and students, time spent on assessment is wasted and learning is not enhanced.

IMPLICATIONS: THE CHALLENGE OF TEACHING ADULT NUMERACY

The ideas presented in this and previous chapters require an expansion of the definition of essential math-related skills to include additional elements such as skills in verbal and written communication about quantitative issues, the interpretation of numerical information encountered in the media, some preparation for making decisions under uncertainty, and the ability and confidence to choose effective modes of response that are suitable for given number-laden problems. Students' attitudes and beliefs about studying and using math in both educational settings as well as in daily, family, and work-related activities should be addressed; these have an impact on motivation to develop mathematical thought processes, to adopt a "mathematical stance" when engaging real-world numeracy tasks, and to continue studying mathematics.

For math education to be effective and meaningful for adults, we must broaden the contexts in which instruction is couched and widen the range of instructional experiences and interactions among students and between students and mathematics content. The instructional principles described earlier imply that numeracy classes may not often resemble the traditional math class in which the teacher makes a presentation while the students watch or write a sequence of computational steps, followed by a "practice" period during which students practice the specific skill just demonstrated until mastery. The process of teaching or guiding adult learners of mathematics requires continuous interaction between the teacher and the student(s) as learning activities are selected, presented, extended, or adapted for students with specific learning difficulties (see chap. 9).

Instruction and activities within an adult math class should focus on "stand-alone" (computational and other) skills and targeted mathematical understanding, but also aim to develop generalized problem-solving, reasoning, and communication skills. By engaging in the problem-solving process and struggling with solutions, students gain the abilities and dispositions needed to apply their numeracy skills appropriately in other problem contexts, as well as develop or practice "component skills" in a meaningful context.

All too often, students work on one skill at a time and are told what algorithms to apply as part of contrived or context-free problems (e.g., "Do all the fraction multiplication problems on this page"). However, as described by Gal (1993; see also chap. 1), in real-life contexts, quantitative or quantifiable elements may be interspersed with other information, and it is seldom specified what to do or what knowledge is relevant. People have to *comprehend* a situation, *decide* what to do, and *choose* the right (mathematical or other) tools from their "tool chest" that will enable them to reach a *reasonable* or effective solution or response. Instructional experiences should thus include a sufficient number of tasks, activities, projects, or problems in which students have to grapple with these aspects of numerate behavior. They should be given opportunities to self-evaluate the quality of their functioning or responses, rather then be told what to do and use the teacher or the textbook as the sole benchmarks for judging the quality of their work.

The framework described within this chapter acknowledges the unique challenges of teaching adults (see chap. 13). To create learning experiences that enhance instruction, teachers and students will need to identify the knowledge—both formal skills and strategies as well as informal intuitions and ways of thinking—that individual students bring to their studies. By designing instruction with this information in mind, teachers will be able to guide students to "construct" their own numeracy learning based on the premise that new knowledge is related to and builds on previous knowledge. In this way, mathematical skills and related literacy skills become cohesive and contribute to the development of "mathematical power" (NCTM, 1989), rather than being a series of separate (and perhaps to the student, unrelated) ideas or skills. Ultimately, instruction should aim to be more obviously pertinent (keeping students involved and coming to class) and more cognitively meaningful (so that students will be more likely to leave with skills that will be retained and can be effectively applied).

A significant goal of instruction concerns helping each student realize his or her own personal goals related to mathematics, which often include passing a high school equivalency examination or a "gatekeeping" test, improving job prospects or skills, or feeling able to support his or her children's math education, among others. It is the ongoing responsibility of the instructor to also identify and communicate to the student goals that he or she may not articulate, or information about gaps in his or her knowledge about which the student may be unaware.

When teachers begin the process of instructional change, they often find it difficult to know where to start. As illustrated in "stories from the trenches," such as those provided in this volume by Arriola (chap. 11) and Curry (chap. 12), changes in educational emphasis require that teachers reevaluate and redefine their roles within both the classroom and the educational process. They often have to embark on a systematic inquiry of the impact of changes on themselves, as well as on their students and program (Cochran-Smith & Lytle, 1995).

Teachers who begin to modify their instructional practices may find it useful to arrange a supportive environment for this change process. They may seek a partner with whom to exchange ideas and get a broader perspective. They may involve students by telling them about new pedagogical ideas and asking for their feedback. They may seek broader and more formal support by forming a local "inquiry circle" or "numeracy group." Groups can take different formats. Some may focus on a single "theoretical issue" of local concern (such as math anxiety), or on one of the instructional principles listed earlier (e.g., how can we link literacy and numeracy instruction?), identifying and reading relevant materials and then discussing implications for practice. Other groups may focus on an instructional topic (such as teaching fractions), trying out ideas and activities and reflecting on lessons learned.

As part of a change process, teachers may feel that challenges are being presented to their own beliefs about the nature of learning math, the nature of math education, how people learn math, how a math class should look, what math skills are important to learn, and what it means to know math (see chap. 14 for some very poignant examples on this last issue). Ultimately, the teacher may no longer view him- or herself, or be viewed by students, as the sole "expert" in the class who knows all the answers as well as the "best" ways to arrive at those answers. Rather, the teacher may increasingly be viewed as a facilitator who participates in the learning process with students by questioning, pushing, explaining, sharing, and also finding his or her own new insights.

REFERENCES

Allan, L., & Lord, S. (1991). *For a number of reasons.* Redfern, NSW, Australia: Adult Literacy Information Office.

Ball, D. L. (1993). Halves, pieces, and twoths: Constructing and using representational contexts in teaching fractions. In T. P. Carpenter, E. Fennema, & T. A. Romberg (Eds.), *Rational numbers: An integration of research* (pp. 157-196). Hillsdale, NJ: Erlbaum.

Bishop, A. J. (1991). Mathematics education in its cultural context. In M. Harris (Ed.), *Schools, mathematics and work* (pp. 29–41). Bristol, PA: Falmer Press.

Brown, S. I., & Walter, M. I. (1990). *The art of problem posing.* Hillsdale, NJ: Erlbaum.

Carnevale, A. P., Gainer, L. J., & Meltzer, A. S. (1990). *Workplace basics: The essential skills employers want.* San Francisco: Jossey Bass.

Cochran-Smith, M., & Lytle, S. (1995). *Inside/outside.* New York: Teachers College Press.

Dees, R. L. (1991). The role of cooperative learning in increasing problem-solving ability in a college remedial course. *Journal for Research in Mathematics Education, 22*, 409–421.

Gal, I. (1993). *Issues and challenges in adult numeracy* (Tech. Rep. TR93-15). Philadelphia: National Center on Adult Literacy, University of Pennsylvania.

Gal, I. (1999). Links between literacy and numeracy. In D. A. Wagner, R. Venezky, & B. Street (Eds.), *Literacy: An international handbook.* Boulder, CO: Westview Press.

Ginsburg, L., Gal, I., & Schuh, A. (1995). *What does "100% juice" mean? Exploring adult learners' informal knowledge of percent* (Tech. Rep. TR95-17). Philadelphia: University of Pennsylvania, National Center on Adult Literacy.

Hembree, R., & Dessart, D. J. (1986). Effects of hand-held calculators in precollege mathematics education: A meta-analysis. *Journal for Research in Mathematics Education, 17*(2), 83-99.

Hiebert, J., & Carpenter, T. P. (1992). Learning and teaching with understanding. In D. A. Grouws (Ed.), *Handbook of research on mathematics teaching and learning* (pp. 65-97). New York: Macmillan.

Hope, J. A. (1986). Mental calculation: Anachronism or basic skill? In H. L. Schoen & M. J. Zweng (Eds.), *Estimation and mental computation* (pp. 45-54). Reston, VA: National Council of Teachers of Mathematics.

Jonassen, D. H. (1996). *Computers in the classroom: Mindtools for critical thinking.* Englewood Cliffs, NJ: Prentice Hall.

Kirsch, I. S., Jungeblut, A., Jenkins, L., & Kolstad, A. (1993). *Adult literacy in America: A first look at the results of the National Adult Literacy Survey.* Washington, DC: National Center for Education Statistics, U.S. Department of Education.

Knowles, M. (1990). *The adult learner: A neglected species.* Houston, TX: Gulf Publishing.

Lampert, M. (1986). Knowing, doing, and teaching multiplication. *Cognition and Instruction, 3*(4), 305–342.

Lesh, R., & Lamon, S. J. (Eds.). (1992). *Assessment of authentic performance in school mathematics.* Washington, DC: American Association for the Advancement of Science.

Manly, M. (1997-98). Calculators in the ABE/GED classroom: Gift or curse? *Adult Learning, 9*(2), 16-17.

Mathematical Sciences Education Board (MSEB). (1993). *Measuring what counts.* Washington, DC: National Academy Press.

McLeod, D. B. (1992). Research on affect in mathematics education: A reconceptualization. In D. A. Grouws (Ed.), *Handbook of research on mathematics teaching and learning* (pp. 575–596). New York: Macmillan.

Mikulecky, L., Albers, P., & Peers, M. (1994). *Literacy transfer: A review of the literature* (Tech. Rep. TR94-05). Philadelphia: National Center on Adult Literacy, University of Pennsylvania.

National Council of Teachers of Mathematics (NCTM). (1989). *Curriculum and evaluation standards for school mathematics.* Reston, VA: National Council of Teachers of Mathematics.

National Council of Teachers of Mathematics (NCTM). (1995). *Assessment standards for school mathematics.* Reston, VA: National Council of Teachers of Mathematics.

Nunes, T., Schliemann, A. D., & Carraher, D. W. (1993). *Street mathematics and school mathematics.* Cambridge, England: Cambridge University Press.

Perry, M. (1991). Learning and transfer: Instructional conditions and conceptual change. *Cognitive Development, 6,* 449-468.

Plunkett, S. (1979). Decomposition and all that rot. *Mathematics in Schools, 8,* 2–5.

Schoen, H., & Zweng, M. (Eds.). (1986). *Estimation and mental computation.* Reston, VA: National Council of Teachers of Mathematics.

Secretary of Labor's Commission on Achieving Necessary Skills (SCANS). (1991). *What work requires of schools: A SCANS report for America 2000.* Washington, DC: U.S. Government Printing Office.

Siegler, R. S. (1991). Strategy choice and strategy discovery. *Learning and Instruction, 1,* 89-102.

Slavin, R. E. (1990). *Cooperative learning: Theory, research and practice.* Englewood Cliffs, NJ: Prentice Hall.

Sowder, J. T., & Wheeler, M. M. (1989). The development of concepts and strategies used in computational estimation. *Journal for Research in Mathematics Education, 20,* 130–146.

Strasser, R., Barr, G., Evans, J., & Wolf, A. (1991). Skills versus understanding. In M. Harris (Ed.), *Schools, mathematics and work* (pp. 158–168). Bristol, PA: Falmer Press.

Tobias, S. (1994). *Overcoming math anxiety.* New York: Norton.

Wearne, D., & Hiebert, J. (1988). Constructing and using meaning for mathematical symbols: The case of decimal fractions. In J. Hiebert & M. Behr (Eds.), *Number concepts and operations in the middle grades* (Vol. 2, pp. 220-235). Reston, VA: National Council of Teachers of Mathematics.

Webb, M. L. (1992). Assessment of students' knowledge of mathematics: Steps toward a theory. In D. A. Grouws (Ed.), *Handbook of research on mathematics teaching and learning* (pp. 661-683). New York: Macmillan.

6

Characteristics of Adult Learners of Mathematics*

James Steele Foerch
Grand Rapids Community Education

This chapter describes the diversity in characteristics of adult learners of mathematics and examines the implications for the individual paths to numeracy that each student must find. It aims to show why adult educators must ask this fundamental question in the singular for each student: "What should this student do in my class?"

---

*Jim Foerch originally wrote this chapter using a "you" form, assuming all readers would be practitioners like him, as he wanted to provoke them to think hard about their own practice. The volume editor found it necessary to use a "we" form in some parts of the chapter to make certain points apply to all those involved in adult numeracy education, be it in teaching, supervisory, administrative, or research capacities. I hope that the editorial process did not inadvertently diminish Jim's strong personal voice. The many case studies sprinkled throughout can be effectively used in staff development activities to promote awareness of diversity and its implications. In chap. 13, Hartman provides additional thoughts about aspects of student diversity that can affect choice of curriculum, teaching methods, instruction, and learning.

INTRODUCTION

Traditionally, mathematics education has been viewed as a single path beginning with the basic operations on whole numbers, continuing through rational numbers, and culminating in a year of algebra taught as abstract symbol manipulation. Subsequently, few students were selected to go on to calculus, and the rest were weeded out. Viewing the mathematics curriculum as a single ladder has been viciously effective at sorting students into groups.

I argue that the single ladder model of mathematics is inappropriate for adult students because it leads either to a mass approach to teaching or to extreme individualization of the curriculum. In the former case, a single ladder approach imposes one generic outcome even though adult students have diverse goals. A standardized, inflexible curriculum implemented through group instruction will be too fast for some students and too slow for others, as it must necessarily ignore each individual's aptitudes and cognitive style.

On the other hand, as students move along the single, narrow path they may become isolated from one another, if teachers who are sensitive to individual differences assume that "they're all at different places and need to do their own work." This isolation overlooks the rich diversity of experience, knowledge, and learning styles that learners bring to the class, and it may fail to make full use of the benefits of learner collaboration and interaction.

Overly individualized instruction may also ignore the reality of the workplace and of everyday life, where learners are expected to work as members of teams and to interact and communicate effectively in and out of their immediate communities. In the technical workplace of today, problem solving is a group effort. It involves listening to others' ideas, sharing one's own knowledge, compromising, and negotiating. Employers prefer applicants who already have such skills. (Carnevale, Gainer, & Meltzer, 1988; Employability Skills Task Force, 1989; Johnston & Packer, 1987; SCANS, 1991). Teamwork skills cannot be taught in isolation; they must be an integral part of every curriculum. The National Council of Teachers of Mathematics has responded to this call by emphasizing group learning, mathematical communication, and cooperative learning in its Curriculum Standards (NCTM, 1989; also see chap. 3 and elsewhere in this volume); many publishers of resource materials and textbooks are now attempting to implement these standards.

To counter the single ladder view, every teacher must answer the question, "What should this student do in my class?," in the singular for each student. This chapter outlines a conceptual framework that can help teachers identify, respect, and utilize individual learners' unique goals, interests, and aptitudes when planning instruction. This framework can also be used with learners to help them identify and articulate their goals and learning preferences. Only when we begin to identify our students as individuals we can create a balance between group instruction and individualized learning and plan instruction that helps them find *their* path to *their* goals. By formally offering diverse paths to numeracy for all students, adult education programs will be able to meet the needs of both the students and society.

The aspects of diversity that we should be aware of are grouped into three separate but interrelated categories: diversity in goals and objectives, in background, and in potential and expected progress. Illustrative composite case studies of learners are presented to remind us that our job is to teach *people*, not merely mathematics. Possible suggestions for teaching are offered in the paragraph entitled "Thoughts" which follows each case description. These thoughts, as well as the general points raised in the main text, are not meant to be prescriptive but simply to serve as possible starting points for discussion about educational experiences that may be required with different learners.

DIVERSITY IN GOALS AND OBJECTIVES

Diversity in goals and objectives is the most important parameter for the numeracy teacher to keep in mind. Each adult has his or her own unique reasons for coming to class, and we must know and respect those reasons to make numeracy learning relevant to each student. Therefore, a basic practice of adult education is to identify these goals, validate them (e.g., through group discussion), and help learners set up markers of progress toward their goals. It is our continual responsibility to remind ourselves, our students, our funding sources, and referring agencies of each individual's objectives. Not all our students can succinctly articulate their goals when they first come to class. But until they can, they will not know where they are headed, how fast or how well they are learning, or the value of numeracy in their lives. Helping each of our students state his or her goals is a prerequisite to learning mathematics because that is the source motivation for learning. (LaMeres, 1990, discusses ways to help students formulate and refine their goals.)

Short-term, Functional Goals

Some adults come to numeracy classes with functional or surface goals. These are immediate, pressing, short term objectives that bring them to our door: Preparing for the high school equivalency GED test, including its mathematics subtest; satisfying the requirements of social welfare agencies; and learning specific job skills or meeting the prerequisites of training programs.

Passing a test. Passing a test such as the GED may be the first step toward the training an unskilled person needs for a better job, better pay, and a better life. That worker may think, "I just need to pass the test." Even though some depth of understanding of mathematics through algebra and geometry is necessary to complete the GED mathematics subtest, the student may say he or she does not have time for that—"Just show me how to do the problem." How do we counter this tendency?

> Patty needs to qualify for admission to some program. She earned mediocre grades in high school mathematics, taking general and business math courses, not college track algebra and geometry. As a schoolgirl, Patty was the kid who often said, "What are we going to use this junk for, anyway?" That attitude kept her from knowing the place of numeracy in real life. On standardized tests she scores at the middle-school level. She may learn well in any mode, but she is often most comfortable as an interpersonal learner and will make your small groups click like clockwork. However, her goal is to move on as quickly as possible, so she sees your class as a hurdle in her path. She wants to know and do the minimum, hence the quality of her work at the adult class suffers.

> Thoughts. You will need to negotiate performance standards for her learning outcomes. Help her connect numeracy with her life and career goals. For example, her description of multiplication and division as used in real life should include partial product multiplication, long division, and estimation; be written in complete sentences; and describe a real situation.

Students who have to complete elementary school material (i.e., are at the pre-GED level in the United States) may not understand how much they have to learn. Indeed, adults with lower math skills often feel strongly that mathematics is just a matter of learning canned algorithms by rote. For students who aim to "just" pass a test, it may be useful to show them that problem solving, not arithmetic, is the skill tested by the GED mathematics test. The teacher should present problems, challenges, and mathematical opportunities which can't be solved by the rote application of algorithms (see NCTM, 1989, and other chapters in this volume, for a discussion of higher order thinking skills in numeracy instruction). With pre-GED students, educators will have to be frank but gentle about the skill level each student is starting from (based on placement test scores) and the need to significantly improve reading abilities to be successful. Teachers will need to go over examples of the reading on any examinations students may have to take.

Terry is a teenager who recently dropped out of school. In the late elementary grades, he produced low to average scores on standardized tests and began getting poor grades. He repeated a grade (class) and was placed in an alternative school; neither of these decisions helped him change his underachieving habits. His stated goal is to get his high school diploma. Because he is young and not academically inclined, the unstated part of that goal is that he wants to get it as easily as possible. Terry has seen too many of his peers "earn" passing grades by simply behaving satisfactorily in school.

Thoughts. For Terry, mathematics is drill and practice on whole number arithmetic, and learning is defined as passing tests (passing is any grade slightly higher than a "fail"). Teach him about demonstrating outcomes verbally, graphically, or with a physical activity. He may not have a predominant cognitive style yet, so have him experiment with different modes of learning. He is young and likes to have a good time. Therefore you will need to be a disciplinarian. Do enforce adult standards of attendance, behavior, and performance. To help him connect mathematics with real life, build his curriculum around community service projects, useful physical tasks in the program, and personal projects.

The mandated student. In many states in the United States, recipients of public assistance and Aid to Dependent Children are required to attend job training classes or adult education centers to remain eligible for their grants. Not all mandated social service clients will welcome the opportunity to become numerate. At first, they will pull out every middle-school trick they recall to keep from doing any real learning. The teacher must carefully balance firm expectations of attendance, punctuality, and progress with a nonjudgmental, accepting atmosphere. As mandated students learn to enjoy learning, teachers can help them formulate more rewarding goals than just staying on welfare. Initial group discussions of "why we are in class" will be gripe sessions. However, if teachers keep asking students to make their own judgments on what numeracy skills they need and use in their lives, they will begin to put their own personal values on learning.

As school-age children, our adult students were told what to do and never heard any relevant reasons why they should learn math or other subjects. When the agency told them they had to attend school to get their money, it was more of the same. But now as adults, they want to be respected and in charge of their own lives. Soliciting each person's opinion on numeracy every week gives them that respect and the power of choice. As they put their own value on learning, they will be relieved to find that there is more to mathematics than drill and practice on isolated skills.

Workplace-related learners. Some students need to learn specific mathematics skills for their job. As working adults with all the responsibilities of family and life, they will be very task oriented. "I need to learn to read a shop rule to the nearest 64th of an inch," or "Just show me SPC (Statistical Process Control)." Employers paying for such training may believe workers only need to learn to apply an algorithm by rote. The best way to respond to requests for superficial learning is to remind the parties involved that a person who truly knows a skill can trouble-shoot and fix problems when they occur. Students should be required to demonstrate their learning by being able to explain why, how, and what they are doing in nonroutine situations.

The goal identification activities mentioned earlier are appropriate for working students. Many workers dream of a better job. Although it may not be part of the job description of some adult teachers to offer career counseling, they can help students make the first step toward realizing a dream by learning to state specific goals.

Long-Term, Fundamental, or Ongoing Goals

Many students have fundamental, long-term goals (behind stated short-term goals, or in addition to them). They may want to become literate and numerate, build their self-esteem, or be able to help their children with homework. The latter is a powerful motivator for parents of school-age children and can be generalized for all students. We should point out often that our basic job on this earth is to take care of children and give them a good life. The better educated we are, the better we can care for children materially and spiritually.

English as a Second Language (ESL) and other adult students report that being able to help their children in school and communicate with their children's teachers are some of the reasons they attend classes (Valentine, 1990). With such parents, we may focus on the mathematics that shows up as their children's homework. Parents from every country expect their children to be taught the same way they were. However, the style and content of K-12 mathematics is changing in many communities. For example, adult students may consider the use of calculators to be "cheating," but their children are required to use them in school. As part of numeracy education, we should be teaching our adults when to use different modes of calculation (estimation, mental arithmetic, pencil and paper, calculator, and computer), to help those parents understand their children's curriculum. Discussing the differences between parents' expectations and their children's experiences is time well spent linking mathematics to life and learning to communicate mathematically.

> Eyob recently immigrated here from Kenya. He learned textbook English in school and completed mathematics courses at the undergraduate (college) level in Kenya. His accent makes it difficult to understand him, and he can only comprehend a small part of what he hears. His goal is to learn colloquial English and American-style mathematics so he can help his children through school. He does not understand why his childrens' teachers spend so much time on cooperative learning, projects, and thematic units because his teachers in Kenya did none of those things.

> Thoughts. Eyob will quickly learn the standard American format for writing math. Include him in all discussions of personal goals and the uses of numeracy in life and work. The format and examples from an adult education class will help him understand his children's curriculum. Enlist the other students to speak with him as much as possible. Provide taperecorded articles with quantitative content. His goals will be to learn to read them aloud and be able to paraphrase the numerical content.

If a numeracy classroom is part of a school system's adult education program, it may also teach students who work over a period of time toward a High School Completion diploma. Typically, adults returning to school feel they are not good at mathematics and have unpleasant memories of math classes. They will be enrolled in the general or business math courses and have no desire to move onto algebra or geometry. To change their perception of mathematics will require more than lectures and kindly advice about the place of numeracy in modern life. Even though they say they "don't know much about math," adult students will have learned a great deal of practical mathematics in their lives. To connect their experiences with the school curriculum, the curriculum department of the school can be asked to provide a list of expected outcomes for each course. If an adult student can demonstrate any of those outcomes, consider giving them credit for those skills, both socially in class and academically in progress records. By using mathematics skill inventories (discussed later) and interviews in class, teachers can focus students' efforts on the skills they need to learn and build their curriculum on such skills. Even skipping one chapter of busy work can be a real boost for adults who remember long hours of drill and practice.

Students themselves can lose track of their long-term goals when they are caught up in the daily complications of dealing with agencies, child care, and so on. In a crowded classroom where the teacher has too many students vying for his or her attention, it is easier to hand out workbooks and tell everyone to start on page one. The students also have a mind-set, a model of the way school is supposed to be. They will be content with a completely individualized curriculum. Yet, overly individualized instruction, as suggested earlier, ignores the reality of workplace and everyday life, where learners are expected to work as members of teams and to interact and communicate effectively in and out of their communities. It is the teacher's job to show students that an integral part of numeracy is learning how it fits into real life. A teacher does this by constantly asking him or herself and his or her students, "Why are you here?" This is an ongoing process of reflecting on goals, reinforcing the usefulness of mathematics, and validating the learners' life experiences and knowledge. Teachers have to show students by example that learning mathematics can be empowering and help them to reach their objectives in life.

DIVERSITY IN BACKGROUND

This section surveys elements in adults' backgrounds that a numeracy teacher needs to know. Certainly students' educational experience is important. But their socioeconomic status, cultural roots, and learning style also may have had a dramatic effect on them.

Consider students' *socioeconomic status*. The least well-off students become distracted by their daily hustle for the necessities of life. They struggle from day to day and have little experience making or achieving long-range goals. On the other hand, some middle-class workers also want numeracy training. They become impatient with their progress because they are used to being productive in the workplace and because they recall the "smart kids" from their childhood. They think they should just be able to "get it" the first time every time. They should be taught that learning is hard work involving challenges, backing up, and relearning along the way.

Socio-conomic status is often correlated with academic skills and habits. Teachers should be prepared to teach students specific study habits, how to reduce distractions, and how to create a learning environment (e.g., turn off the television, turn on the lights, plan activities to keep the children busy.) No one is born knowing the techniques of learning, and many families do not have a scholarly tradition to fall back on. Students should be shown how to experiment with study techniques to find what works best for them.

According to Gardner (1993) and Lazear (1991), we all operate with *multiple intelligences* and learn in several modes with one or two predominant ones. The seven intelligences as analyzed by Gardner (1993) are listed here with mathematical examples:

- Verbal/Linguistic   Discuss a problem
- Logical/Mathematical   Inductive reasoning, "seeing" the solution
- Visual/Spatial   Make a model of the problem
- Body/Kinesthetic   Use body parts to estimate measurements
- Musical/Rhythmic   Recite multiplication facts to the melody of "Twinkle Little Star"
- Interpersonal   Solve problems as part of a group
- Intrapersonal   Describe what is unique about your path to mastery of trigonometry.

The first time our students return to class as adults, they try to learn the way they thought they were supposed to as children. This means they try to operate only in the verbal/linguistic and logical/mathematical modes. Numeracy educators may inadvertently reinforce this narrow, restrictive view of intelligence by relying too heavily on published materials. Lazear's and Gardner's work shows that a person learns best when taught in several of the modes corresponding to the seven intelligences. The prescription is to learn about multiple intelligences and give your students permission to learn in the modes best suited to them. For instance, body/kinesthetic learners will fidget in their seats, shaking their legs and tapping their pencils distractedly until the teacher points out to them that they should be using their whole bodies to learn. Suggest that they work at the chalk board using large characters, or that they study while standing up at a high table; if possible, teachers should use gross motor, room-sized projects to make lessons more concrete.

Cultural differences in learning style are related to multiple intelligences. Although I have found the individual differences within ethnic and racial groups to be far greater in range and significance than the differences between groups (Dunn & Griggs, 1990), in some communities it may be politically expedient to talk in terms of cultural differences. For example, some writers suggest that Native Americans' predominant mode is intrapersonal learning (Wall & Arden, 1990). In that case, instruction should emphasize each individual's path to mastery of mathematics and it is unique use in life. Parenthetically, that will be validating and reinforcing for non-Native Americans as well.

> Tiffany is a young mother. She dropped out of school to have her baby after a disappointing school career. She does poorly on standardized tests, has low self-esteem, and feels isolated from the community because she has to take care of her child all the time. School may be her only social outlet, making it doubly important in her life. Tiffany's goal is to be able to earn a decent living for her child. However, it will take more time than she anticipates to bring her basic language and mathematics skills up to speed.
>
> Thoughts. She is convinced she is not good at math because she could not understand story problems or fractions. The only way to counteract her math phobia is to give her opportunities for success. Explain gently and firmly that unless she does achieve success now, she will not be able to handle the postsecondary training she needs for herself and her child. Reinforce often the importance of her role as a good mother and her decision to return to school. Discuss with her the need for quantitative thinking as she cares for her child. Better yet, if she brings her child to your daycare center,you can help her measure and record her baby's growth and development. She will readily perceive the value of numeracy in her main role in life—being a good mother. Because she is young, Tiffany should experiment with different modes of learning to find her own. Continually demonstrate the connections between basic numeracy skills and her future career.

Our students vary in educational background. In basic skills programs, most students will have dropped out of school during their teen years. As teachers establish rapport with them, they will find that some were actually pushed out of school—the girls to have their babies, the boys because they were growing up too slowly, or because the mental, emotional, and social challenges to learning were too great. Most have regrets about having wasted youthful opportunity and many will express anger at the schools or at themselves. Teachers should reflect those emotions back with their best active listening techniques and then show their students by example that an adult numeracy program is not the place to rail about the inadequacies of the K-12 system, but rather a time to capitalize on the changes in their live that brought them back to school. Accentuate the positive.

Significant numbers of adult students have learning or developmental disabilities (see chap. 9). Your intake procedure should identify these challenges because they contributed to failure in the past and will affect goals, progress, and methodology now. The next section provides a direction for dealing with these difficult, sensitive issues by teaching the students to set and periodically revise realistic goals.

Finally, teachers will meet adults who are college bound or preparing for the skilled trades. They will fly through the program and be impatient with delays. Adult education centers must be structured so that these people can move on to the next phase of their education as soon as they are ready (perhaps by using relevant technology; see chap. 8).

> Michelle is college bound. She speaks and writes well and scores high on standardized tests. She is in your numeracy class because she dropped out of school. She was a quick learner, so the low-track, general math classes she was enrolled in seemed boring and irrelevant to her. But now Michelle has a reason for formal education; she is ready to work on a career. Because she has a definite goal, she can generalize from texts to career situations. Her goals are both ambitious and realistic, so she does high-quality work. This is the student for whom mathematics texts are written because her cognitive styles are verbal/linguistic and logical/mathematical. Enjoy helping her.

DIVERSITY IN POTENTIAL AND IN EXPECTED PROGRESS

In the previous section I discussed aspects of students' background, including prior educational experience, socioeconomic status, cultural roots, and learning style, that have implications for the rate and extent to which they can achieve their short- and long-term goals. Teachers have the delicate task of judging adult learners' potential and helping them choose *attainable outcomes*. I emphasize both words because each adult learner has his or her own unique set of learning goals (see earlier section) and perhaps numeracy outcomes; when they first enter class, some students will have unrealistic timelines for themselves. There may be a woman who cannot read the local newspaper who will tell the teacher she expects to complete high school by the end of the semester. There may be students who complete chapter after chapter in math workbooks but cannot understand how a checking account works.

We should keep in mind and teach students that *an attainable outcome will lead to a demonstration or performance of math skills in the near future.* If our instruction and the student's activities do not end with such a demonstration in a reasonable length of time, learning has not occurred. Students will then have to rewrite their goals. The teacher will have to change the learning activities and situation. This process may be threatening or disappointing for the student, but we should remember that outcomes must be attainable in the learner's time-frame.

As mentioned earlier, our poorest students live from day to day. The weeks it takes to complete a whole curriculum can seem like forever to such students. Adults with learning disabilities or other learning challenges (see chap. 9) may never be able to complete a published curriculum. The teacher must help impatient, slower learners set short-term goals. For example, every basic mathematics student should be able to add with "regrouping" of single numbers into tens. But for students who do not even have the basic facts of addition memorized, a suitable outcome for one week might be to demonstrate regrouping with base-10 blocks for numbers up to 20.

Mary has a developmental disability. As a child she was placed in special education where she enjoyed mastering algorithms for whole number arithmetic. She learned to solve story problems by trying the four operations in turn with the numbers presented because she could not read the problems. In her special education classes there were few opportunities to use mathematics in real-life situations, so Mary views numbers and arithmetic as strictly a school activity. She scores at the middle elementary grade level or lower on standardized tests. She is socially adept, joins in activities enthusiastically, and is a reliable follower. Mary often learns psychomotor tasks very quickly because she is a kinesthetic learner. As an adult student she works hard and enjoys workbooks and computer-aided instruction, but is unable to transfer her math skills to real-life situations.

Thoughts. Mary may live in an adult foster care home or on her own. She may work in a sheltered workshop, but is dependent on supplemental security income and other public assistance. She does not foresee any changes in her employment status or lifestyle. Yet, if you ask her why she wants to go to school, she will tell you she wants a high school diploma. She does not realize that her mediocre reading ability precludes work at the high school level. Just as she sees math as repetitive drill, for Mary reading is a matter of filling in blanks in workbooks in school. Her comprehension is well below her decoding and phonics abilities. Mary's curriculum should be based on real tasks and community service projects that improve her life, her family, and the community.

In earlier sections I discussed the most important parameters of diversity among adults of which numeracy teachers must be aware. That said, the practitioner may find the need to add other dimensions to the list and should consider the caveats in the next section, which discusses assessment issues are discussed.

ASSESSMENT TOOLS AND PROCESSES

When students, the general public, and funding agencies think of assessment, they think primarily or only about paper-and-pencil, standardized tests (especially in the United States). Testing instruments such as the TABE, WRAT, Silveroni, or Key Math, discussed later, aim to give an accurate view of people's academic skills.

- TABE   Test of Adult Basic Education. Reports math and reading skills by strand and satisfies government testing requirements.
- WRAT   Wide Range Achievement Test. Quickly gives general academic skill level.
- Silveroli   Especially useful for adults with very low reading ability.
- Key Math   Lengthy, individually administered math test that gives the teacher the opportunity to observe the learner's work habits and problem-solving skills.

It is crucial to constantly remind students, funding agencies, and the general public that, although each standardized achievement test has its own focus and fits a somewhat different testing situation, *none* of these tests reflects potential success or failure in life, and all have serious limitations (see chap. 15). Students should be helped to understand that standardized assessment instruments can help teachers place the students in the program, and perhaps show general progress in learning (as well as satisfy funding agencies). However, the techniques of Outcome Based Education, wherein each learner becomes responsible for demonstrating his or her learning, are more appropriate (Brandt, 1993; Spady & Marshall, 1991).

Through intake interviews, application forms, and school transcripts when available, each student's educational history can be collected. Look for a sudden drop in grades or grades repeated. These may have led to alternative school placements, dropping or being pushed out, or special education placements. Not only do these suggest challenges to learning, but they are often symptomatic of negative attitudes toward learning and low self-esteem on the part of the learner. (Should your program use standardized tests that report test scores as "school grade equivalents," scores lower than 6.0 indicate students who may have significant challenges to learning). Van Groenestijn (chap. 16) describes other aspects of initial assessment pertaining to mathematical thinking strategies.

Find out your students' socioeconomic status. Ask each learner how he or she lives and works now, where he or she would like to be in the future, and how he or she plans to get there. Listen for realistic steps and timelines. Some adults may not plan to change the circumstances of their lives. You will have to help these individuals create numeracy goals and objectives that are appropriate for their situation.

The teacher can identify students' primary and secondary learning styles by teaching them about multiple intelligences (Lazear, 1992) and helping them analyze their own behavior (also see LaMeres, 1990, for learning style inventories and their use in class).

All the information you have collected will inform your impression of the student. That first impression, be it positive or negative, consciously chosen or intuitively felt, accurate or wrong, will guide your initial instructional decisions for him or her. However, goal setting and assessment should be ongoing, weekly activities. As the teacher speaks and works with the students individually and in groups, he or she recreates his or her impressions of them even as they learn to recreate their own self-images. By setting, adjusting, and finally achieving their goals they can become more numerate adults. The formal assessments mentioned earlier will show certain aspects of achievement in mathematics.

However, adults who become truly numerate will be able to assess themselves and show what they have learned.

The danger here is that our observations and assessment, even taking into account all the sources of variation mentioned so far, do not fully characterize a person's potential to learn or to reach his or her goals. The teacher has to judge each student's potential on a continuing basis. Anything less is to prejudge them and limit their future. Noting progress through demonstrations of new skills in real life situations, and redefining and setting both surface and fundamental goals, is an ongoing process that a teacher must build into class time each week. Only by constantly communicating with students can teachers make accurate judgments about their potential.

OTHER ADULT LEARNER ARCHETYPES

Throughout this chapter I have described some typical adult learners. Others whom one may meet in the numeracy classroom include the following students:

David has a mental illness and is required (mandated) to attend your program by his mental health agency or the corrections system. He may score high or low on standardized tests. His predominant learning style may be in any mode, but the mental illness interferes with his learning. As a child he was labeled emotionally impaired. He has no particular plans or view of the future and so sees little utility in numeracy. His goal is day-to-day maintenance.

Thoughts. Work with him and his caseworker to identify and control symptomatic behaviors. David enjoys mathematics workbooks because they are safe and do not require interaction with people. To reinforce tiny steps toward participation, include him in all group activities even though he may choose to be a passive observer.

Alice has been a homemaker for 15 years. She wants or needs to upgrade her skills because she is about to enter the workforce. She feels anxious about going back to school, afraid she will not be able to keep up. Alice is a perfect example of a person who has math phobia: she recalls math as a mysterious subject that made her feel dumb.

Thoughts. Find out how she uses mathematics as a home-maker in budgeting, cooking, and tracking her children's growth. Alice does have considerable quantitative ability. She will appreciate discussions relating textbook math to real life and learning to state her knowledge in formal terms (see also chap. 14 about the mathematics implicit in women's tradition-al work, such as needlework, weaving, or knitting). Alice's goal is to get a job. She knows she needs math, but does not have an inkling of what mathematics skills different jobs require. Help her to explore different careers and to see examples of numeracy in the workplace.

Frank has a learning disability. He is employed and is a steady, reliable worker. Abstract symbolization, which is involved in so much of school work, has been challenging for him all his life. His standardized test scores are low, and he was placed in a "resource room" as a child. He comes to your class wanting to improve his mathematics skills, already understanding his learning disability.

Thoughts. Frank is an excellent kinesthetic or visual/spatial learner; show him the moves once and he's got it. Because Frank's goal, to improve his math, is general and undefined, set short-term performance outcomes. An example would be for him to learn to use a calculator to calculate the mileage his car gets. He will learn to do this and other quantitative tasks that pertain to real life. He will not show much progress on standardized tests.

Mike is learning new job skills. His standardized test scores are high and he has quite specific goals. "I need to know sines, secants and tangents for my CAD/CAM (Computer Assisted Drawing/Computer Assisted Machining) program." He is a strong verbal/linguistic or logical/mathematical learner and so can use texts or computer-assisted instruction inde-pendently. He assesses his own progress and does quality work. However, he will only be in your class as long as neces-sary to reach his immediate goals. Show him rigorous stan-dards for demonstrating mastery and give him as much tech-nical information as he can handle.

The last case study is Ned. Ned is a "normal" adult learner and, as such, does not exist. Every one of our adult students has his or her story to tell and deserves a chance to tell it. Each person has challenges and barriers to overcome from past life, and each person has something unique to offer employers, the community, and the world. Teachers should help them discover their special talents; that is what will ultimately justify all their hard work to become (more) numerate.

IMPLICATIONS

The descriptions just given have been based on the students this writer has been privileged to work with. The characteristics of adult learners have been presented in this way to emphasize that we teach human beings in all their many dimensions.

Numeracy may be defined as the mathematics an individual needs to live a successful life. Because the adults who come to numeracy classes do so from different backgrounds, for different reasons, and with different abilities and potentials, each one needs his or her own unique mathematics. This view of numeracy as many paths to many goals is not new: It is what Maria Montessori spoke of so many decades ago. Piaget taught us how each child constructs reality (and knowledge) for him or herself. It is part and parcel of John Dewey's practical education. The teaching and supervision journals of today are speaking of *authentic learning* and *constructivist* approaches in which each learner creates meaning and invents his or her own algorithms (see chap. 2, this volume). By balancing individual and group instruction, by helping each learner chart a path to his or her goals, and by teaching students to assess their own potential and accomplishments, teachers can help each one become (more) numerate.

Because our students are adults, they need to know we value and respect them, their goals, and their life experiences. Instruction that reflects students' goals and draws on their experiences is not only the antidote to an inflexible, canned curriculum, but a way of demonstrating the important applications that math has in everyday and workplace contexts. As with all disciplines, mathematics is only meaningful in context.

Numeracy education is one part of a path to better life. Adults do not have time to waste—neither their own nor the teacher's—on mere arithmetic. By keeping in mind the parameters of diversity of adult learners, we (and you, the teacher) can help them learn the numeracy skills they need. By remembering how unique all students are, teachers will be able to help them along their own paths.

REFERENCES

Brandt, R. (1993). On outcome-based education: A conversation with Bill Spady. *Educational Leadership, 50*(4), 66-70.

Carnevale, A., Gainer, L., & Meltzer, A. (1988). *Workplace basics: The skills that employers want.* San Francisco: Jossey-Bass.

Dunn, R., & Griggs, S. A. (1990). Research on the learning style characteristics of selected racial and ethnic groups. *Journal of Reading, Writing, and Learning Disabilities International, 6*(3), 261-280.

Employability Skills Task Force (1989). *Employability skills profile: Progress report of the Governor's Commission on Jobs and Economic Development and the Michigan State Board of Education.* Lansing, MI: State Department of Education.

Gardner, H. (1993). *Multiple intelligences: The theory in practice.* New York: Basic Books.

Johnston, W. B., & Packer, A. B. (1987). *Workforce 2000: Work and workers for the 21st century.* Indianapolis: Hudson Institute.

LaMeres, C. (1990). *The winner's circle: Yes, I can!* Newport Beach, CA: LaMeres Lifestyles Unlimited.

Lazear, D. (1991). *Seven ways of knowing.* Palatine, IL: Skylight Publishing.

Lazear, D. (1992). *Seven ways of teaching.* Palatine, IL: Skylight Publishing.

National Council of Teachers of Mathematics (NCTM) (1989). *Curriculum and evaluation standards for school mathematics.* Reston, VA: Author.

Secretary's Commission on Achieving the Necessary Skills (SCANS). (1991). *What work requires of schools: A SCANS report for America 2000.* Washington, DC: U. S. Government Printing Office.

Spady, W., & Marshall, K. C. (1991). Beyond traditional outcome-based education. *Educational Leadership, 49*(67), 67-72.

Valentine, T. (1990). *What motivates non-English-speaking adults to participate in the federal English-as-a-second-language program? Research on adult basic education, vol. 2.* Des Moines: Iowa State Department of Education.

Wall, S., & Arden, H. (1990). *Wisdomkeepers.* Hillsboro, OR: Beyond Words Publishing.

7

Adult Numeracy at the Elementary Level: Addition and Subtraction up to 100

Wim Matthijsse
CINOP

This chapter explores the development of a pedagogy for working with numbers up to 100 within the framework of an elementary numeracy course for adults. The goal of this approach is to enable school mathematics to help adults cope with daily life by strengthening the informal and natural strategies adults already have and use.

INTRODUCTION

Numeracy courses in adult basic education (ABE) in the Netherlands are still very new, as adult education was almost nonexistent in this country until the early 1980s. Because of this short history, concepts, materials, and ideas are still in the process of development. Many of the students in ABE have learned some numerical skills at school, but their knowledge is often patchy. On the one hand, they do not know how to use this knowledge in practical daily life situations; in other words, their numerical knowledge is not always functional. On the other hand, they have developed their own natural informal strategies, which are often context-bound, to cope with numerical problems in daily life. Thus, there is a serious gap between "school mathematics" and the mathematics that adults use in daily life.

This is seen especially in the target groups for the elementary level: illiterate and semi-literate people. Many of these students are immigrants who did not go to school in their homelands. But even illiterate people have much knowledge at their disposal, although their knowledge is often limited and bound to specific situations. It could be said that their mathematics is somewhat fossilized. We have tried to search for a pedagogy (some may prefer the term "andragogy"), specifically for mental models, that allows for the use of context-bound strategies while helping students develop more formal algorithms they can apply to a wider range of situations.

There are also adults, both immigrants and native inhabitants, who have gone to school for a few years. Usually, their school careers were fragmented, and they did not benefit much from the experience. These adults often do not have very positive memories of school mathematics. They have an inaccurate but explicable view of mathematics as mostly incomprehensible rules, and they feel that mathematics has nothing to do with daily life. They feel that in daily life they learn by doing things and solving real problems, but at school they learned rules and had to do rote calculations.

With this in mind, this chapter explores the development of a pedagogy for working with numbers up to 100 within the framework of an elementary numeracy course for adults. Our goal is to teach mathematics so as to help adults cope with daily life by strengthening the informal and natural strategies they already have.

In adult basic education (ABE) in the Netherlands we distinguish among three levels of adult numeracy courses: an elementary level, a higher fundamental level, and a currency level that prepares ABE students for vocational training. In this chapter, I focus only on the elementary level, for which the focus is presently restricted to:

- Addition and subtraction up to 100
- Multiplication tables
- Mental arithmetic
- Decimals that occur as the results of measurements (money, length, weight, etc.)
- Practical measuring (money, time, length, weight)
- Elementary spatial orientation

Note that standard column arithmetic, fractions, and percents are not included at this level. Specifically, the chapter concentrates on addition and subtraction up to 20 and then to 100, addressing the following questions:

1. *What informal methods do adults use in daily life? Why is it important to know about them when starting instruction?* I address several numerical problems that adults encounter in their daily lives and describe certain kinds of informal strategies they use to add and subtract. I discuss why it is important to start with students' informal strategies, and then describe a pedagogical framework for a numeracy course at the elementary level, using a functional-realistic approach.

2. *What mental models can we use for instruction in addition and subtraction? How can we base formal algorithms on informal mental operations?* I suggest some instructional methods for teaching addition and subtraction up to 100 and propose a strongly oral and interactive curriculum that focuses on a real and rich understanding of numbers, the use of structured counting, the use of the money-board as a mental model for memorization of addition and subtraction facts up to 20, and the use of the empty number line for addition and subtraction up to 100. I also suggest some ways to use the empty number line to support a more natural transition from mental arithmetic to standard column arithmetic and present examples of students' activities and working materials to illustrate these concepts.

WHAT INFORMAL METHODS DO ADULTS USE IN DAILY LIFE TO SOLVE FUNCTIONAL PROBLEMS?

When developing an elementary numeracy course for adults, a very important starting point is the use of everyday life situations as a source for investigation. Here is one example:

You find the flyer from a supermarket on the bus. What are the prices? What are the bargains? What are you going to buy? Why? Do you make a list or do you just go to the supermarket? How much money do you take with you? Do you pay with exact change or with large bills? Can you estimate the cost? Do you check the receipt or do you decide not to because there is a long line behind you?

In the Netherlands we have made such lists of so-called functional situations described from a numerical point of view. Of course, the relative importance of various situations will differ for each student, but every adult must cope with some of them. These kinds of problems require that students be able to perform basic operations up to 100. They particularly have to be able to add and to subtract.

Research done in the Netherlands on problem solving has found that illiterate people demonstrated quite different patterns of solution strategies from adults who had been to school. For instance, van Groenestijn (1993, and see chap. 16) found that illiterate people in the ABE classes used several forms of "nonstandard" or invented mental arithmetic for handy calculating. To illustrate this point, next are two examples of solutions devised by an illiterate student to the functional problems in Figures 7.1 and 7.2, followed later by response from a semi-literate student to the problem in Figure 7.3. The quotations are the students' own verbalizations of their solution processes.

(Note: The examples in this chapter use Dutch notations for writing monetary and numerical values which may be different from the conventions used in some other countries. An example is the use of comma instead of a period (or a space) to represent thousands or decimal portions of numbers. Both chaps. 15 and 16 mention such topics among others when discussing assessment issues involved in evaluating knowledge of immigrant students who have to use a new notational system in their new country).

Analysis of these and similar examples reveals a number of interesting points. The student uses benchmarks in his arithmetic, often round numbers that he can easily handle. He also makes use of characteristics of the monetary system such as bills (notes) of f10,- and f100,-. If he has to determine the difference between two amounts of money he does this by counting in certain steps and by addition, but not by subtraction. His procedure depends on the numbers, and he demonstrates that he has confidence in his own algorithm.

Figure 7.3 depicts the work of a semi-literate student who went to school in her childhood. (Note the initial error in summing "15" and "17" into "33"; this sum is then assumed correct and used further on the right side of the figure).

In this case, the student was torturing herself by searching, quite unsuccessfully, for the "right formula." With semi-literate people who have had some schooling, we often encounter this search for the right method, usually algorithms that are poorly understood. They stare at the numbers while asking themselves, "What is the sum?" Their school-learned mathematics seems to prevent them from dealing with numbers in the more natural way the illiterate person used in the previous exam-

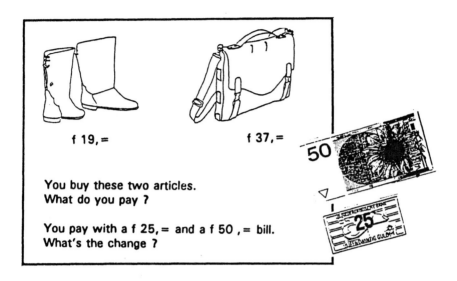

Figure 7.1. What's the change?

Solution:
> "19 and 1 makes 20. 20 and 36 makes 56 (total cost)."
> "From 60 to 75 makes 15. From 60 to 56 makes 4 (by 56+4 = 60).
> Together that's 19 (amount of change)." (Note the student uses
> 60 as a benchmark).

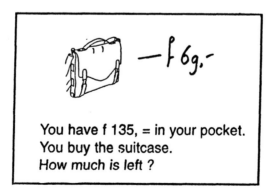

Figure 7.2. How much is left?

Solution:
> "From 100 to 135 makes 35. From 100 to 69 makes 31 (by
> 69+1+30) Together 66."

Figure 7.3. Column arithmetic

ples. It is also quite obvious that the student was not very confident with her method. (See the end of chap. 16 for a dialogue with another semi-literate student that illustrates the same difficulties with a poorly under-stood algorithm and lack of confidence.)

What was the "right formula?" She was looking for the traditional algorithm of column arithmetic (again, assuming 33 is the right sum).

$$
\begin{array}{r}
50 \\
- 33 \\
\hline
17
\end{array}
$$

This algorithm is quite involved and assumes several internal steps. Had she been able to access the proper algorithm, she would have had to (recognize the need to) mentally split up the 50 into tens and ones (she has five "tens" and zero "ones") as well as the 33 (three "tens" and three "ones"). In the next step, the "ones" must be subtracted from each other, so the student must try to subtract three from zero. Because that is impossible, she must borrow a "ten" from the five "tens," so she can subtract three from ten. The result is seven. She now has four "tens," so she subtracts three from four to make one. Therefore the answer is 17.

This algorithm is usually taught to students with concrete manipulatives (such as blocks, cubes, or rods) that enable the teacher to address issues of position of digits and value of numbers. However, we find that many students have particular difficulty with subtraction using column arithmetic, for two reasons.

- They have to memorize and organize a lot of things in their head. Of course, they may work with concrete materials like blocks, but at some point they have to be able to do the process abstractly. They often make mistakes by forgetting to carry or borrow, and they have problems in subtracting from zero. They also often become confused and mix up the tens and ones.
- Many problems that arise from real-life applications are not recognizable as subtraction. The illiterate student in the first example would probably start with 33 and add seven and ten, not recognizing the problem as subtraction at all. (The student may think of the problem as a "difference" problem, a context in which it may be easier to go up from the lower number to the higher number, than to go down from the higher number to the lower one to find the difference between two numbers.)

Thus, the procedure for standard column subtraction does not allow for the use of context-bound and informal strategies of adults. Often one sees students looking at these problems in a forced, artificial way: "How do you do this calculation? What is the rule?" They do this because in traditional mathematics education, strict rules are taught for solving these kinds of problems.

These examples are typical of what we have seen in ABE in the Netherlands. In their daily lives, adults have their own ways of solving problems and their own context-bound strategies, especially when dealing with money. This knowledge is not learned in school but through daily life activities and is well integrated within general learning processes containing a strong social-emotional component. If adults have learned some mathematics at school, the rules that they learned there often "block off" and frustrate the more natural and informal ways of problem solving that they are able to use.

We have recognized and built on these experiences through an alternative pedagogy in which we give formal column arithmetic a less important place in favor of mental arithmetic. Before I discuss this in more detail, I present a more general framework that guides the development of the teaching of ABE mathematics in the Netherlands.

A PEDAGOGICAL FRAMEWORK

This section outlines the general functional-realistic framework for ABE mathematics education in the Netherlands, which is built on the work of the Realistic School (Gravemeijer, 1989; Streefland, 1989; chap. 16, this volume). The Realistic Mathematics Education (RME) approach was developed beginning in the 1970s in an effort to improve mathematics education in the primary schools. At this point over 80% of the primary schools in the Netherlands have adopted this approach to mathematics. In general, RME involves:

1. *Functional contexts and applications.* It is important to have a pedagogy that will systematically work within daily-life situations and that uses realistic contexts, models, and procedures that allow for the application of context-bound and informal strategies within these contexts. Continuously working with practical applications, especially money, is helpful because all adults deal with money every day. There are three reasons for this approach:
   a. functional situations can be used to introduce relevant numeracy problems which motivate students;
   b. thinking about daily situations can help students when they are calculating; For instance, thinking of money can help when a student sees 2.45;
   c. contexts can be used to help students assess the outcomes of procedures they have learned in classroom lessons. For example, a student may be comfortable in school with a sum of 514 obtained by applying a learned algorithm to the computation of 26 + 38, but the same student in a store may demand a recalculation if asked to pay $5.14 for two items priced at $0.26 and $0.38.

2. *Useful mental models.* It is also important that students formalize their knowledge and skills so that their numerical skills can be applied to a wide range of situations and utilized in an efficient and flexible way. For that purpose mental and calculation models are used to bridge the gap between real-life context-bound skills and more formal mathematics (such as the empty number line described later). Students need thinking models that will not overrule their own meaningful context-bound arithmetical strategies by replacing them with a "superior" formal system. These models should also allow students to gradually formalize their insight and informal knowledge. Finally, the models must enable the student to construct knowledge and skills on his or her own. This is based on the notion that students should construct and reconstruct their own knowledge and skills rather than reproduce the teacher's way of doing things (see chap. 2 for further discussion of constructivism).

3. *Students' own productions and constructions.* Students solve problems in different ways and at different levels. That is a fact. In our pedagogy we encourage this diversity because it makes students aware of the richness of mathematics and makes them realize that mathematics is not a one-dimensional activity. The pedagogy must allow and encourage unique student solutions that can then be shared with other students.

4. *Interaction and verbalization.* In this process of self-construction of mathematical skills and insights, it is very important for a number of reasons that students have the opportunity to share their numerical and arithmetic strategies with their fellow students. First, this encourages reflection on their own solutions and those of others, and leads to the sharing of handy strategies. Second, explication and verbalization help to develop mathematical thinking. Third, verbalization gives the teacher an insight into the way the student thinks and calculates. If necessary, the teacher can then diagnose problems or discover missing links. Mathematics should be a very social activity, contrary to the popular image that it has.

5. *Interrelationships between the different parts of mathematics.* Most of the parts and subjects of mathematics are connected. For example, subtraction is linked to addition and number sense is linked with measurement. The pedagogy must help students to become aware of these linkages and to use them in applications. Overall, we do not aim for a pedagogy in which the student is learning algorithms by mimicking the teacher and then trying to apply these learned algorithms. Rather we start with all kinds of realistic situations, from which the students construct their own mathematics by using mental or thinking models, by reflecting on their own mathematics, and through interaction with fellow students and teachers. Last but not least, the education is focused on having fun while doing mathematics and on developing a good mathematical attitude.

INSTRUCTIONAL METHODS

Before describing instructional methods of addition and subtraction up to 20 and up to 100, I clarify three assumptions that guide our instructional program.

1. *The meaning of numbers.* Numbers are not just numerals. Numbers can be an amount (13 students in a group), a number

in a row (bus line 7), a measurement (5.6 miles or $10), an iden-
tification number (bank account no. 1978426; a telephone number), or the result of a calculation (5 + 7). Students must learn
to recognize the meaning of the numbers in context. The reference points they create become the language of the numbers.
Especially at the elementary level, a number by itself only has
meaning if the student can think of a context for it. We must
help the students to "dress" the numbers, or to create a rich
environment for numbers, as in the next two problems:

- You are in a waiting room in the hospital. You have number
  13; number 7 is being helped at the moment. How many
  clients are ahead of you?
- You are in a group of 13 students of which 7 are female.
  How many students are male?

In the first situation you might count "8, 9, 10, 11, 12, 13," and in
the second you might do the calculation "13 minus 7;" the strategy one
uses depends on the connection between the numbers and the context. A
teacher can develop students' appreciation for this concept by regularly
presenting bare numbers and asking them if they can name a situation
in which those numbers might occur, or by presenting situations and
asking them if they can think of numbers that might occur in that context. It can be said that a real understanding of numbers encompasses all
the situations in which numbers are used, or, in other words, the adult
phenomenology of numbers. This kind of understanding is necessary if
one wants to be able to link the contexts of numbers and the strategies
for the basic operations with numbers and thus to achieve transfer
between contexts.[1]

---

[1]The author refers here to what might be construed as part of the broad notion of
"number sense," yet raises a fundamental issue that is often neglected in most
discussions of number sense, that of the context or origin of numbers, and the
influence of a person's understanding of that context on the way in which that
person makes sense of numbers. When we expect students to demonstrate
"number sense," such as by knowing what a "billion" or a "0.01% chance" may
mean, such numbers cannot be evaluated in a vacuum. People's subjective
understanding of processes and events in the world, that is, their phenomeno-
logical view of reality, have to be considered (and enhanced as part of the teaching process) if we want students to reason effectively about numbers they
encounter in realistic contexts. One also has to make sure that a student's understanding of the context of a problem, and of the origin of given numbers, is consistent with that of the teacher or of other students.)

2. *Working with numbers up to 20 and up to 100 is not the same.* We distinguish addition and subtraction up to 20 from addition and subtraction up to 100 because there is a fundamental difference between the two: students have to memorize all the addition and subtraction facts up to 20, but that is impossible with larger numbers. Calculating up to 100 is accomplished by means of strategies and procedures. However, students need to memorize most of the addition and subtraction facts up to 20 before they learn to calculate up to 100.

3. *Verbal instruction methods.* Because the particular mathematical language of symbols is difficult, and many of the illiterate or even semi-literate students have problems with written instruction, there is a strong emphasis on verbal and visual education.

ADDITION AND SUBTRACTION UP TO 20

Although skill in addition and subtraction of numbers up to 20 ultimately relies on memorized number facts, we emphasize counting and modeling with a "money board" (described later) as a tool to help students understand the basis of the operations and make concrete the meaning of the numbers. "Counting" includes a number of related strategies that reflect different levels of familiarity and confidence with the addition and subtraction processes. Both counting and the use of the money board encourage students to decompose numbers into convenient component parts, an approach that helps support addition and subtraction with larger numbers and useful informal mathematics for daily life.

Counting

In traditional mathematics education, counting has a bad image: counting is forbidden one is are expected to calculate without one's fingers. Many adults are ashamed to use their fingers and often work with their fingers under the table. As a result of the investigations of the Realistic School in the Netherlands, researchers discovered that counting, and especially *structured counting*, is a necessary and important part of learning the basic operations. Counting is also a goal in itself because it plays an important role in everyday life: adults often have to count things and understand numbers that are the result of counting.

Being able to count is of great importance for calculating. Because many adults can already count and, more importantly, use counting as a natural strategy to solve numerical problems (as in the examples in the previous parts of this chapter), counting must be regarded as a first step toward addition and subtraction.

For example, teachers can let students count an amount of coins as shown in Figure 7.4. It is quite interesting and instructive to see how students (and also teachers) count. There are all kinds of strategies including counting one-by-one or in pairs, doubling one row, two rows, or three rows, and multiplication. By allowing students to tell each other their counting strategies, they can also learn from each other.

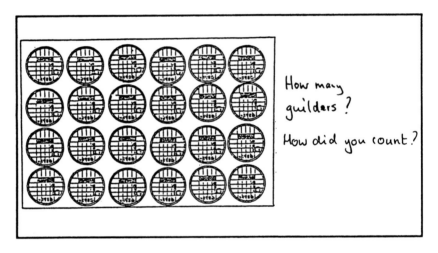

Figure 7.4. Array of guilders

There are at least three important levels of counting strategies. These are:

1. *Resultative counting:* At this level, students are able to count amounts of things one-by-one.
2. *Structured counting:* At this level, students can count in a structured way, such as counting in steps of 2 (doubles), 5, or 10; if two amounts are joined, counting further by starting from the biggest amount ("counting on"); and, if a part is taken away, counting back from the starting amount.
3. *Abstract counting:* The next step is counting amounts when one cannot see all the objects. This is the symbolic level, in which the number "6" can be seen as representing the "hidden" amount of six objects. At this moment the student has developed mental pictures of numbers. The student sees at one glance structured amounts and does not need to count all six dots on the dice. He or she sees at one glance the number "6."

Structured counting can be taught by giving students either structured amounts (as in the previous example of counting coins), or nonstructured amounts (such as a handful of paper clips). Counting nonstructured amounts is more difficult because students have to make the structure by themselves. There are a lot of different structuring strategies that students can discuss with each other. This is also an appropriate time to talk to students about how to organize everyday counting problems, such as counting the spectators at a soccer game, the students at school, the money in their pocket, and so on.

Working with dice can help students memorize in a natural way many elementary strategies for addition, subtraction, and creating number pairs up to 12. In addition, dot charts with structured patterns of pairs of dots (doubles) and with the five-structure (groups of fives) have been found to be helpful. Both are explained next.

Addition Up to 20 with Dice and Money Board

We first start with addition and subtraction up to 10 or 12. Students have to memorize all of the basic addition and subtraction facts with numbers (or amounts) up to 10. Dice can be used for that purpose in a very natural way.

At first the student starts with one die and then moves on to two dice. Gradually, the student will know many or all of the possible number combinations by heart. One can proceed even further by making up riddles, such as: "Someone rolled a seven. If you can see one die with so many dots, how many dots must be on the other die?" One can also roll the dice and ask what numbers are on the opposite sides of the dice. At that moment the teacher can introduce the symbolic language $(5 + 4 = ?)$ and start with the next step: addition and subtraction up to 20.

In traditional school mathematics education in the Netherlands, the rule of "making tens" is applied when a sum will exceed 10. (This rule means that $7 + 6$ will be calculated as $7 + 3 + 3$; one always has to start by "filling up" to 10 and then adding the additional amount). Typically, no informal working methods are allowed, yet if one asks students how they add numbers up to 20 the answer is that they use all kinds of informal counting methods in addition to "making tens," such as:

- Doubling (6+6 = 12) or almost doubling (6+7 = 12+1 = 13)
- Working with the five-structure (6+7 = 5+1+5+2 = 10+3 = 13)
- Counting on (9+2 = 9+1+1 = 11)
- Splitting evenly (6+8 = 7+7 = 14)
- Combinations of these methods

It is important to let the students verbalize and share their strategies. This exercise can be placed within a real context by asking questions, such as: "If you buy two articles (items) that cost $6 and $8, what do you pay? What is the change from a $20 bill?"

In some of the strategies mentioned, the importance of the number "five" should be recognized. This is because one can see five or fewer than five objects all at once; one does not have to count that amount one by one. (Try it by counting on fingers: eight fingers is seen at once as 5 and 3.)

In looking for a model based on five and allowing for the use of different computational strategies, we use the money board with a five-structure. The student represents numbers on the board by placing into holes in the board either single (loose) guilders (each representing one unit), or fitting over the holes a "five-strip" (containing a drawing or a fake/real coin of 5 guilders, and representing the amount five). For numbers greater than four the student must use the five-guilder strips. (Note: Guilders are the standard currency in the Netherlands. In January 2002 guilders will be replaced by the Euro.)

The money board supports both the use of doubles and five-structures for counting, and thus for adding and subtracting. Figure 7.5 illustrates the two main strategies of working on the money board using the example "6 + 7". On the left hand of the board, the student makes "six" by fitting a single guilder on the top and a five-strip on the bottom. Likewise, the number "seven" is created on the right hand side with two singles on top and a five-strip on the bottom. The student immediately sees two five-guilder strips and three single guilders, so in his or her mind the student thinks:

6+7 ==> 5+1+5+2 = 10+3

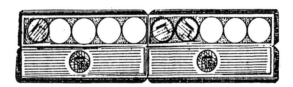

Figure 7.5. Money board

Alternatively, the student can solve "6+7" by filling up to 10. As illustrated in Figure 7.6, the student creates (makes) the number six by putting a five-strip and one guilder on the bottom row. Next he or she "fills up" the row to 10 with four guilders ("borrowed" from the seven) by putting down a second five-strip over the bottom row that has the single guilder and then putting the three remaining guilders in the upper row.

$$6+7 ==> 6+4+3 = 10+3$$

Figure 7.6. Filling boards to 10

Subtraction Up to 20

We have observed that most adults have little difficulty with addition of numbers over 10 (up to 20) but do have problems with subtraction of numbers over 10. Two basic informal subtraction strategies that are often used are "counting back" (i.e., starting with the larger of two given numbers and going "down" the number line until reaching the smaller number) and "adding on" (i.e., starting with the smaller number and finding how much to add to it to reach the larger number). However, as mentioned earlier, traditional school instruction does not connect carrying and borrowing algorithms to these informal solution methods. It is often assumed that weaker students continue to count because they do not know how to split up numbers into tens by heart, because their working memory does not function properly, or because they feel insecure with abbreviated methods of counting. The pedagogy must connect instruction to the informal working methods of the students. Also, an attempt must be made to make more sophisticated procedures accessible to weaker pupils who persist in counting, and as an indirect result, may irrevocably fall behind in advanced arithmetic.

By viewing subtraction as the inverse of addition and by making use of students' knowledge of the addition facts up to 20, we take advantage of students' informal counting strategies. We encourage students to relate subtraction to addition. This can obviously be achieved by using the money board and enabling students to structure the problem as they want (i.e., starting either from the larger or smaller numbers). In using the money board to solve "13-8", for example, a student can represent 13 as two five-strips and three single guilders. The student will take away three single guilders and then one five-strip, leaving one five-strip as the answer.

We make sure to also provide verbal feedback that makes explicit the strategy used when working visually on the money board and that reiterates the verbal representation of the problem (from the point of view of the solver as determined by, e.g., watching the student work on the money board). For example (our feedback is written on the right),

$12 - 9 = 3$ "correct, because 9 and 3 makes 12" (addition strategy)
$11 - 2 = 8$ "wrong, because 11 minus 1, minus 1, makes 9" (subtraction strategy)

We have seen that if the difference between the two numbers in a subtraction problem is small, students tend to add or to count on. However, if the difference between the numbers is large, they will subtract. We call this pattern of choosing a computational procedure the "two-sided strategy of subtraction" (we see this pattern again later when discussing subtraction up to 100).

To focus attention on this pattern of strategy choice, the teacher can consciously give students subtraction problems with large and small differences. It is important to discuss these two strategies for subtraction with the students. As students become aware of different strategies for subtraction they tend to use and apply them more thoughtfully.

Benefits of Using the Money Board

We have found that by using the money board model, counting is not suppressed but rather mastered. Use of the money board seems to support students' development of conceptional understanding of number, addition, and subtraction by providing a familiar, concrete model whose parts can be manipulated in meaningful ways. The range of the ways students use the money board include:

1. Perceiving and making number pictures

   - Quickly reading concrete pictures of numbers represented on the board
   - Quickly making concrete pictures on the board (by using five-guilder strips for numbers above four)
   - Working with flash cards
   - Exercises with covering, such as giving students a quick look at the board, then covering it and asking, "What do you see?"

2. Working on the board by showing the guilders and verbalizing, using addition strategies:

   - The strategy of working with doubles, calculating with fives
   - The strategy of "filling up," or making a row empty/full to 10

3. Looking at the board without touching it. (The board gives visual support for students' thinking, and the 20-structure functions as a mental model.)

4. Mentally visualizing addition and subtraction operations on the board to provide conceptual support and confidence.

ADDITION AND SUBTRACTION UP TO 100

The Number as a Measurement

When we look at numbers as tens and ones we focus on the place value aspect of numbers. This aspect plays an important role in traditional column arithmetic because one has to be able to split up numbers into tens and ones. But in our approach, early in the curriculum we pay more attention to the ordinal aspect of the number system, that is, the number as it appears in a series of numbers. This ordinal aspect is not only very important for a pedagogy for arithmetic up to 100 because it is very closely linked to structured counting, but for mental arithmetic and estimating with larger numbers. This approach is illustrated with the tape measure in Figure 7.7.

Students must learn to give numbers a place on the number line, sometimes without having access to full information. In everyday life, there are many situations that require this skill, including reading measuring instruments, finding house numbers on a street, and working with counters.

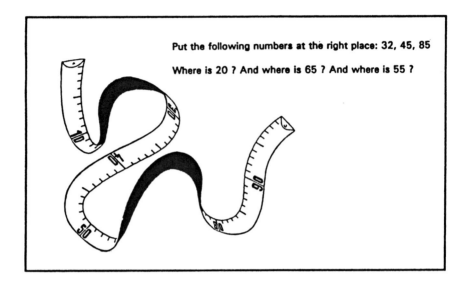

Put the following numbers at the right place: 32, 45, 85

Where is 20 ? And where is 65 ? And where is 55 ?

Figure 7.7. Measuring tape

Working With the Curtain-Rail

Students explore the ordinal aspect of numbers by using a variety of forms of a number line, starting with a filled number line. For that purpose we make use of a device we have developed that we call the *curtain-rail*. This device is a long rod on which there are 100 sliding units. The units are in two colors (red and black), with the color changing every ten units, as shown in Figure 7.8:

Figure 7.8. A curtain-rail

The idea behind this device is that students can manipulate the units while counting, by sliding them to the left or to the right on the rod (in a way reminiscent of an abacus). The following are some class activities that use the curtain-rail:

- Show the number 37 by sliding 37 units to the left. (After making the number "89", students may see that they can also make "89" by sliding 11 to the right.) Ask questions, such as, "Does 37 lie closer to 30 or 40?"
- Make 37, add 10, and again add 10, and so on. What numbers do you pass? You can also do the same with steps of 10 backward.
- Make multiple steps of two units forward and backward. Are there patterns? Make steps of five.
- Make 37, go 19 forward and then go 19 backward. Where do you end up?
- How many units are there between 37 and 62?

While working on solutions to problems such as this last subtraction problem (numbers between 37 and 62), students discover they can determine the difference between two numbers in two ways, by counting on from 37 in steps of three, 20, and two up to 62, or by starting from 62 and taking away 37 units (adding on and subtracting strategies). If the task is finding the difference between 7 and 62, a student will often count back from 62 in steps of 2 and 5, or set 62 and then slide back 7 units.

By asking the students which method they find the easiest and why, they will become aware that the efficiency of a particular strategy may depend on the given numbers. Our investigations have shown that at first students solved all the problems by subtraction, but after working some time with the curtain-rails they changed to "adding on" in situations like "62 - 47" or "73 - 56", realizing that solutions can be found in fewer steps. It is also very important to let the students verbalize their activities and strategies as they work with the curtain-rail.

The most important activities with the concrete curtain-rail are positioning numbers on the number line and discovering relations between numbers and operations. Although the curtain-rail helps students explore the two-sided character of subtraction (counting up and subtracting), students should learn to think in jumps and not just in the one-by-one counting that is supported by the sliding units.

Working With a Representation of the Curtain-Rail and an Empty Number Line

After students feel comfortable and competent with the curtain-rail, we move to a more abstract representation; color segments are painted on a curtain-rail but numbers are not present and there are no sliding rings. Finally, students work with an empty number line. We use the same activities as with the real curtain-rail, but the students have to draw a line and position the numbers from the problem on the line. Then, instead of sliding units with their hands, they draw the jumps they make.

Longer or shorter arithmetic methods can be illustrated on an empty number line, including counting, abbreviated counting (i.e., 27 as "+3, +20, +4") or via efficient arithmetic (27 as "+30 - 3"). Figure 7.9 provides three examples of students' use of the number line to solve subtraction problems. The problems being solved are 76 - 48 and 56 - 25.

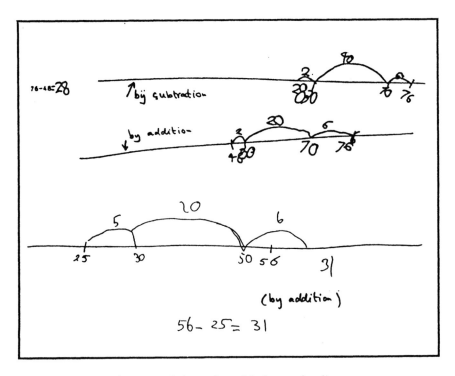

Figure 7.9. Subtraction with the number line

Figure 7.10 shows a student's solution process to the functional problem 43 - 27 (presented as a word problem as in the picture). The work on the left shows the student's solution attempt at the beginning of the course; she then tried column arithmetic and made a classic mistake. On the right, is her strategy after working with the curtain-rail and the number line; she solved the problem making use of the empty number line by adding on.

In working with the empty number line students naturally abbreviate their counting by increasingly using fewer steps. This process is enhanced by letting the students continuously discuss among themselves how they do their calculations and jumps. Eventually, students will not need the drawn line anymore because they will have the line in their minds as a mental model. Overall, our investigations show that students often solve subtraction problems like this very differently from traditional algorithms if you give them the "space" to do it in their own way.

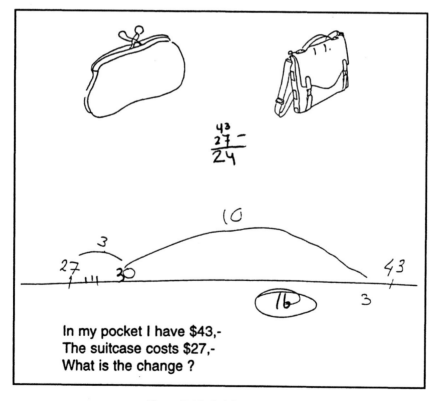

Figure 7.10. Solving 43 - 27

Mental Arithmetic Methods by Columns

In teaching column arithmetic we do not aim directly for written arithmetic based on standard algorithms. Informal solutions are also allowed as predecessors or supplements to standard column arithmetic. Figure 7.11 shows three alternative, permissible column-based approaches to the solution of "64 - 37."

Each of these column methods is supported with concrete materials and mental models designed to encourage flexibility in decomposing numbers. Both the ordinal as well as the place value aspect of the number is acknowledged.

In the functional-realistic pedagogy we use, addition and subtraction up to 100 are not immediately routinized, leaving room for various mental strategies of arithmetic. We begin by building on counting, and therefore, the arithmetic is ordinal. Then, somewhat later, we emphasize the place value aspect of arithmetic by reinforcing the usefulness of the number 10 through concrete materials such as the curtain-rail but not with the standard addition and subtraction algorithms.

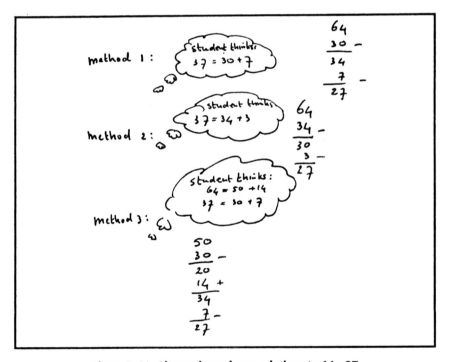

Figure 7.11. Alternative column solutions to 64 - 37

SUMMARY AND IMPLICATIONS

Problems that students have with standard column arithmetic, and especially with subtraction, occur because these algorithms do not allow for the use of context-bound, informal strategies. The money board and the curtain-rail provide concrete models that allow students to perform counting strategies and "adding on" strategies. We need these useful manipulatives and models to bridge the gap between the informal context-bound strategies that adults have and formal arithmetic.

The empty number line is also a useful model for the development of arithmetic understanding, procedures, and computational shortcuts. The relation between the operations of addition and subtraction (and later between multiplication and division) is heavily supported by the number line.

What role does the calculator play in this curriculum? In our opinion the pocket calculator plays a very important role in ABE but should not be used for elementary addition and subtraction. A responsible use of the pocket calculator is only possible if a student can estimate to check answers. For estimating and mental arithmetic, it is important to understand, among other things, addition and subtraction up to 100. Therefore, we believe calculating up to 100 must be done mentally and not with a calculator.

Each of the described strategies and models has marked consequences for memorizing and making arithmetic skills automatic. They can facilitate memorization of the basic addition and subtraction facts as well as allow students to move forward from their own natural starting points.

REFERENCES

Gravemeijer, K. (1989). The important role of context problems in realistic mathematics instruction. In C. A. Maher, G. A. Goldin, & R. B. Davis (Eds.), *Proceedings of the Eleventh Psychology of Mathematics Education-North American Chapter Conference* (Vol. II, pp. 125-128). New Brunswick, NJ: Rutgers University.

Streefland, L. (1989). Realistic mathematics education (RME): What does it mean? In C. A. Maher, G. A. Goldin, & R. B. Davis (Eds.), *Proceedings of the Eleventh Psychology of Mathematics Education-North American Chapter Conference* (Vol. II, pp. 121-124). New Brunswick, NJ: Rutgers University.

van Groenestijn, M. (1993). *Moet ik uit het hoofd of mag ik rekenen?* [An analysis of mathematical strategies of 20 Moroccan adults in adult basic education]. Utrecht, The Netherlands: University of Utrecht.

8

Technology and the Development of Mathematical Skills in Adult Learners

Betty Hurley Lawrence
Empire State College

For many reasons, technology offers great promise for improving the learning of mathematics by adult students. (The term "technology" as used here relates to a wide range of tools, from simple and widely available ones such as audiotapes and videotapes, to computer-based and other advanced tools.) In an age in which problem solving using quantitative information has become a necessity for many decisions required of adults, and when electronic technologies are increasingly being used in all facets of modern life, adult educators need to consider carefully how best to integrate use of technology into the teaching-learning process.

This chapter is aimed at adult educators who are presently using little or no technology and who are unsure as to what extent technological tools should or could be incorporated into their practice. The chapter focuses on the following five questions:

1. What are some key reasons for incorporating technology into work with adult students who are studying mathematics?
2. What benefit or impact on learning can different technological tools offer?
3. What are some criteria for assessing the effectiveness of a technological tool?
4. What types of technological tools are available in the area of mathematics learning?
5. How do I start?

WHY USE TECHNOLOGY IN ADULT EDUCATION?

Educational technology that can be used to support the learning of mathematics includes many forms. This chapter focuses on several key types of technological tools: videotapes, audiotapes, calculators, computer software, and telecommunication networks.

An obvious rationale for using technology in adult education is that technology surrounds us. As adults, we are faced with technology on a daily basis—as we try to get cash from an Automatic Teller Machine, use an telephone answering machine or a videotape, bet at a racetrack using computerized bidding machines, trust the supermarket that the right prices of products can be found through their bar codes and summed without visible human application of mathematical computations, or allow the supermarket cashier to accept a credit card for groceries. It would be less natural to ignore technology in our classrooms rather than embrace it. The SCANS report (1991) identifies the importance of familiarity with technology for a competitive workforce. For those adults who are uncomfortable with technology in general, its use in educational contexts can provide an opportunity to increase their comfort level. Of course, there are other advantages, many of which are identified by Turner (1993) in the NCAL report, *Literacy and Machines: An Overview of the Use of Technology in Adult Literacy Programs.*

Technology provides adults with the privacy they sometimes desire, especially when working on basic skills that they are embarrassed about not possessing. Technology can provide the learner with more control over his or her learning, because the learner can take charge of what and how learning will occur. Regarding the teaching process, technology increases the opportunity for individualization in instruction as well as for quick feedback and a report on progress. In addition, technology can increase flexibility because it ideally can be made available during all hours of the day.

Despite repeated calls for increasing the use of educational technology, given its obvious benefits, it appears that only a minority of adult educators, at least in the United States, use technology in their instruction on an ongoing and systematic basis. In a 1993 assessment, it was found that no more than 15% of literacy providers use computers regularly for instruction, and many do not use them at all (Office of Technology Assessment, 1993). A more recent survey (NCAL, 1994) found an increase in use, although access and use is still a problem. A survey that examined practices in adult numeracy education in the United States (Gal & Schuh, 1994) found that, although over 80% of programs surveyed reported availability of some computer software for teaching mathematics, less than 25% of learners gained regular access to this software.

It is certainly not the intent of this chapter to advocate the use of technology with all learners all of the time. Rather, the goal is to reflect on when it is appropriate to use technology, with whom, and for what purposes. The key is to integrate technology into practice in ways that will improve the learning experience.

TECHNOLOGY AND LEARNING OF MATHEMATICS

The danger in writing about the use of technology is that it may feed the temptation to divorce technology from its application and perceive it as an end in itself. Technology should be viewed as a means to meet an educational goal. Separated from this goal, it becomes ineffective at best and sometimes damaging.

With this in mind, technology holds much promise for the enriching of mathematics education in general and for adult students in particular. In an age in which facts have decreasing value and knowledge changes at an alarming rate, students must be prepared to draw actively on mathematics as a tool to investigate and solve real-life problems (Grandgenett, 1991). As described by the National Council of Teachers of Mathematics (1993; also chap. 3, this volume), "knowing mathematics is doing mathematics" (p. 7). Technology should (and, we argue, can) help students do mathematics and learn to think mathematically.

The assumed goal for the use of technology in mathematics education is to facilitate the learning of mathematics. By using technology, we expect that the learner of mathematics will have a deeper, fuller understanding of the concepts involved as well as an increased ability to apply diverse methods to solve a range of problems. Levesque (1989) maintains that the use of computers in classrooms motivates and empowers learners and increases literacy. Through the use of a computer or calculator as a computational tool, one is able to work on "messy" problems while still focusing on the problem-solving strategy.

Computer programs, videos, and new multimedia options, such as interactive video, can provide the learner with powerful images to aid in understanding mathematical concepts that are often visual in nature.

ASSESSING THE VALUE OF A TECHNOLOGICAL TOOL

Because learning mathematics is the primary goal, criteria have been established by the author to identify which technological tools will enhance the learning process.

Active vs. passive learning. Does the technological tool foster an active approach to learning? Certainly the pressing of keys on a keyboard or pad indicates some form of activity, but is it the type of activity that motivates the student to construct his or her own understanding of mathematical concepts? Some adults returning to school expect to learn while being passive, so it is especially important that technological aids are used in ways that dispel this myth.

Developing critical thinking and problem solving skills. Does the tool focus on single steps that are well-defined or is the student encouraged to work through a multistep, open-ended process that requires use of critical thinking and decision making? How much choice is the user given and are multiple answers allowed? For example, are 3/4 and .75 and 0.750 all accepted? Can the tool be used as a problem-solving tool, or is it just taking the learner through predetermined routes? Technological tools that are flexible in a way that utilizes an adult's natural problem-solving skills drawn from experience will be especially effective. Used appropriately, a tool should assist the adult learner with building bridges between life skills and an academic approach, which many adults find to be quite foreign (see chap. 7).

Opportunities for cooperative learning. Does use of the technology allow for cooperative learning? Although there is a place for independent use of any tool, cooperative work offers many benefits (Johnson & Smith, 1993) and can enhance the learning of mathematical skills, as well as help to develop mathematical communication skills. Technology that permits working with others can take advantage of these added benefits. Cooperative learning is sometimes difficult for adult learners, especially if they are feeling unsure of their knowledge in mathematics. Yet, some may have worked in teams at their workplace. The key would be to set up a comfortable learning environment where errors are expected and process is at least as important as product.

Added value. What does the technology offer that a math workbook on the same subject matter would not? Is this a technology that primarily provides text for the learner to read or only routine exercises to solve? These capabilities may sometimes be important (e.g., for learners who need some practice in a form more suitable for their learning preferences; see chap. 9 regarding students with specific learning difficulties or with perceptual or language limitations). Yet, added value should be examined. For example, is the user given a chance to experience a visual representation (perhaps even a three-dimensional view) or the opportunity to go through a simulation to get a better understanding of some mathematical concepts? Can the tool be a bridge between abstract mathematical ideas and real-world events?

Learner control. An advantage of technology that was mentioned earlier is that it can provide learners with more control over their learning situation (Turner, 1993). Does the technology under consideration increase learners' control over the learning environment? If so, how? For example, does the learner have options as to how to proceed through the learning experience or is there only one set path? Does the learner control the pace? To many adult learners, control of their learning is an important issue and one that is often ignored in the hierarchical, teacher-centered classroom. Technology can often move more control over to the learner. Teachers will need to monitor student progress because "goal-oriented high achievers are more likely to be motivated by learner control than low achievers" (Steinberg, 1991, p. 138).

In the next sections, simple technological resources (audiotape, videotape, and calculators), computer software, and telecommunications and networked resources are reviewed in light of the criteria identified earlier.

SIMPLE TECHNOLOGY RESOURCES

Audiotape

Because of the visual nature of mathematics, one may initially think that audiotapes would not be appropriate. However, if we approach mathematics as a language, we can begin to see some benefit in talking and writing about mathematics. Connolly and Vilardi (1989) discuss ways to incorporate writing into the mathematics curriculum. (See also chap. 10 about process writing, and chap. 12 about journal writing, for examples of ways to enhance communication in the mathematics classroom.) For some adults, it is important that they hear mathematical terms. How many students struggle with mathematical terms and pronunciation? To assist the auditory learner, an audiotape made by the instructor can be very helpful. The content of this tape could be a discussion of vocabulary, but could also include the recording of examples of quantitative information, such as advertisements from the radio. As an assignment, a student may prepare an audiotape in which he or she discusses mathematical concepts in his or her own words.

Advantages of this application are certainly cost and availability. Although not "high tech," tape players are inexpensive and currently more widely available than computers. Tapes can be played at home as well as in the car or while taking a walk. A disadvantage, of course, is the lack of a visual component, although the tape can accompany a text or worksheet with graphics included.

In terms of the five criteria listed earlier, audiotapes fare well in the area of added value because they add an auditory component that is not provided by the written word. The level of activity would vary, with the highest level connected to the student's creation of a videotape. Active listening is certainly desirable, but passive listening may also occur. Similarly, the use of an audiotape may or may not enhance critical thinking. Learner control certainly exists through the ability to use the on-off button as well as fast forward and rewind, but audiotapes do have a linear quality to them.

Videotape

Videotapes have become a valuable resource, especially because there is ready access to a videotape player in the United States and many other industrial countries. Three categories of videotapes are discussed: instructional videos that are commercially available, videos created by students, and videos created by a teacher or workplace trainer for students.

Numerous videos and video series have been developed in recent years, and new ones are being produced to enhance the learning of mathematics. Examples are videos created as part of the Annenberg project (e.g., For All Practical Purposes, Against All Odds) or by KET-Kentucky Education Television in the area of GED preparation. Through these and many other high-quality videotapes, connections are made to real-world applications of mathematics through on-site interviews and observations. The "field trips" provided by these tapes are a powerful way to demonstrate the relevance of mathematics. Less sophisticated tapes (but also less costly) are now often available from textbook publishers. In most cases, these are of the author or other educators and mostly review sample problems. Many adult learners find these helpful, especially if they can bring them home for viewing. There also are some videos that are not classified as instructional but contain mathematical information.

In elementary and secondary schools, students have been encouraged to create their own tapes. Adults can certainly do this as well and many already have the equipment. For example, they could record their teaching of a mathematical concept to another, or videotape a mathematical activity at home (e.g., cooking, home repairs, planning a trip or a party).

Of course, teachers could create their own tapes as well, such as to show an interview with a local applied mathematician or with a company representative discussing math needs in the workplace, or to showcase students who have successfully completed a course or unit on mathematics. A related category of videos is one that is not meant to teach mathematics but to stimulate a discussion by teachers and students alike about general aspects of learning or "doing" mathematics, such as math avoidance or anxiety and how to address it during the learning process.

The advantages of videotapes are clear. High availability is obvious. They are also easy to use and can be used at home, a definite advantage for many adult students. They provide audio for learners who learn by listening as well as video for visual learners. They certainly can enrich the learning environment through the addition of mathematics in motion. The major disadvantages are that viewing a videotape can be a very linear and passive activity, and the pace may be too fast for some learners. A more active approach to learning can (and perhaps should) be encouraged by recommending that the student take notes and stop the tape to answer questions on an accompanying worksheet. The videotape is not conducive in a natural way to collaborative learning but can be incorporated as a resource in a collaborative situation (e.g., viewing a selected portion and discussing it in the classroom). Learner control of the tape is possible, as with the audiotape, but the creator of the tape executes the highest control over the content, and the linear nature of the tape restricts learner control. A video can provide good information for the problem solver but does not inherently enhance critical thinking.

Calculators

The hand-held calculator is a powerful tool that has often been underutilized. Because it reduces the drudgery of calculation, it can provide increased opportunity to do problems that require extensive calculations. Ideally, this means that a student can focus on solving the problem, rather than get buried in calculations.

The calculator can be used for more than messy computations, however. With a calculator, one can discover patterns. The book *Family Math* (Stenmark, Thompson, & Cassey, 1986) contains many interesting and simple activities that can be done with calculators with the added advantage that they are activities that family members can do together. Examples for games include "Aunt Bebe's Costly Calculations Contest," "Calculator Paths," "The Lost Rules," and "Pseudo-Monopoly." In addition, many math journals have a section on calculator activities.

Because calculators do differ, it may be wise to have a number of the same kind available. Some of the major distributors, like Texas Instruments, may have a calculator loan program, and many will give free training workshops. It is important that the calculator is appropriate for the task, with the least number of buttons needed being optimal.

The graphical calculator contains a small screen that displays graphs and contains functions to enter and manipulate data, produce graphs or plots and zoom in on interesting sections. It should be seriously considered for students taking algebra and advanced courses. The opportunity to see graphs so easily and manipulate them as well as the ease of assigning values to variables is impressive. Textbooks now exist that incorporate the use of a graphical calculator throughout a math course.

Advantages include low cost and portability. Adults are often less anxious about using calculators than using computers, and they like having something that they can hold in their hand. And, as stated earlier, the calculator permits the student to focus more on solving the problem, rather than getting buried in numbers.

Disadvantages and challenges include dealing with the many types of calculators that students may possess. In addition, it is not wise to assume that students know how to use their calculator. A discussion of what the keys can do is often necessary, especially keys such as the memory key. The calculator is not easily read by more than one person at a time, but this only means that roles need to be assigned in collaborative learning so that one person is the technologist and others in the group become a leader or recorder.

However, there has been concern expressed about the calculator weakening computational skills. It is certainly not uncommon to see a person become overly dependent on a calculator and/or accept outrageously incorrect answers from a calculator. Yet, studies exist that indicate that use of a calculator can enhance computational skills as well as improve problem-solving skills and attitudes toward mathematics (Hembree & Dessart, 1992). Certainly, a calculator can be used to assist in developing estimation skills.

One can also see the calculator as an opportunity to improve error-checking skills. A software package, such as What Do You Do

With a Broken Calculator (Sunburst/Wings for Learning), can be used in conjunction with a calculator to develop activities that enhance error-checking skills. Furthermore, students can first solve a math problem mentally and then use the calculator to verify their calculations.

COMPUTER SOFTWARE

The purpose of this section is not to identify all the individual packages that are available, in part because new and improved ones are released at a rapid pace that would make any attempt at a comprehensive review obsolete, but mainly to identify types of software to look for, depending on the reader's educational goals. There is some discussion about minimum hardware requirements.

Although computers such as the Apple IIs, Atari, or the BBC in the United Kingdom, are still available, the focus is mainly on IBM-compatible computers (DOS and Windows) and on the Macintosh. For the IBM-compatible, Windows-based programs have become increasingly popular. The Windows environment, which now offers a graphical interface and mouse usage similar to that found with a Macintosh, simplifies computer usage for the unfamiliar computer user. Most new software packages are being developed for this environment.

Software packages have been placed into the following categories: instructional, simulation, quantitative analysis, and integrated (obviously, some packages fit into more than one category).

Instructional Types

Although all computer programs could be considered instructional, certain features enhance their instructional value. Several types are considered in this section, from drill and practice to instructional games to programs that include text as well as problems to work on. Media include floppy disks as well as CD-ROMs and videodiscs.

Drill and practice. The earliest types of computer software for mathematics instruction were for drill and practice. The format of these is familiar: a student is given a series of problems to solve and is told whether answers are correct or not. There are certainly advantages to these programs. The student is given the opportunity to practice certain types of problems and skills to gain computational (and emotional) confidence. Feedback is immediately available and it is given privately, preventing embarrassment. In some situations, students can be paired. The dialogue between partners that ensues about the activity can greatly enhance the learning of each participant. In addition, the use of a drill and practice program can free teacher time to work with students on more creative activities and/or to work individually with students.

Newer programs have improved capability in providing feedback or guidance to students. Some programs branch off to different sections depending on the student response, provide hints if requested, or return students to an instructional mode (as opposed to "testing" mode) if too many errors are detected. Some drill and practice programs that are based on advanced logic (and described as being Intelligent Tutoring Systems or based on Artificial Intelligence principles) can analyze some student errors and offer the student or the teacher corrective actions; some of them incorporate enough graphics or text in their feedback and remedial branching to considerably elevate their instructional value.

The potential for self-assessment cannot be ignored. Through the use of a computer-based self-assessment package, a student can become aware of his or her strengths as well as areas that need strengthening. This shifts the responsibility for control of learning to the student and encourages the development of lifelong-learning skills. In mathematics, in which adults are more likely to trust the judgment of the instructor rather than their own (see chap. 11 for Arriola's account of this process), this transfer of responsibility is an important outcome.

Of course, there are disadvantages to drill and practice programs. Often, skills are practiced in isolation, with little connection between and among skills, or between the computer-based practice and the textbook's focus. How helpful is this for the transfer of skills to similar situations and to more complex or realistic problems? Often, there is inflexibility in terms of what answers are accepted as correct, which can be very misleading to the user. Many programs still perpetuate the myth that mathematics deals with a right/wrong world where ambiguity is not allowed. Recent writers have argued against this approach to mathematics learning (Buerk, 1985).

Multimedia. The term *multimedia* is being used quite a bit recently. This category basically includes systems that use more than one medium to develop concepts, typically with text, graphics, animation, and sound. Multimedia programs, if cleverly designed (which is not always the case), can help accommodate diverse learning styles of different learners. CD-ROMs and videodiscs are common resources that provide sufficient storage for these media. However, it is unclear whether videodiscs will survive because of their expense and their replacement by larger capacity CD-ROMs. Both CD-ROMs and videodiscs have the advantages of hypertext, which means that they can be searched quickly enough so that a user need not go from beginning to end in a linear fashion. Rather, one can jump to different sections of the disk depending on what is needed.

The videodisc looks like a record, except that it is read digitally by a special machine. Currently, it can contain much more information than a CD-ROM disk, especially in terms of video. It can also be tied to a computer software package that eases the searching capability, and it can build exercises around the information available on the videodisc. Because of the expense of a videodisc (it requires a videodisc player, a computer, either a special monitor or two monitors), this technology has not become widely used in the classroom.

CD-ROMs are becoming increasingly prevalent and faster, and many inexpensive computers now come with a CD-ROM drive as part of a full multimedia equipment package. CD-ROM disks, which can currently only be read from Read Only Memory (ROM), are being developed into recordable CDs that can be written onto like floppy disks. As this development proceeds, currently expensive prices for recordable CDs are coming down, and adult education programs may even be able to produce custom-made multimedia for local use.

Multimedia programs have much potential in terms of graphical representation and integration of visual and audio information. Imagine a program that shows the terms of an equation actually moving to their solution. Or, consider being able to move to an interview with someone who actually uses an application just covered on the screen. The richness of the multimedia environment holds much promise for the adult learner of mathematics. Learner control is increased as the user can jump to points of interest easily, rather than plod linearly through the material. Different learning styles can be adapted to, including visual and auditory. Connections can quickly be made to "real-world" problems through the integration of full-motion video material. Furthermore, these programs can use the computer hard drive to keep a record of student progress throughout the material, so that an instructor can identify problem areas and make recommendations for further study.

Computer Games. There have been a good number of mathematical games created, primarily for the K-12 market. Some of these games are quite creative and help users develop skills in logic and basic problem solving. The growing use of CD-ROM disks has enriched the environment these games can provide through the increased use of graphics, animation and sound. These games need to be reviewed carefully to see if they are appropriate for a larger adult audience. Those that are not geared specifically for young boys (e.g., sports-based games) have the greatest probability of being appropriate for adult students, especially women. In any case, even if an adult educator does not find certain computer games appropriate for adult learners of mathematics, they can still be recommended to adult learners who have access to computers at home and who seek ways to enhance their children's learning of mathematics.

Simulation

Simulation programs create a model of a realistic situation and enable the learner to see the influence of different actions on this situation. For example, one can simulate space travel, a lengthy trip, a chemical reaction, a production process, or how a machine operates. Most of the development of educational programs using simulation has been in the K-12 area. Examples include integrative programs such as Oregon Trail and the Voyage of the Mimi. In Oregon Trail, characters share their exciting stories about travel or living along the Oregon trail, which was used by American pioneers on their way to the West Coast. The user has the opportunity to choose among dozens of different sites to travel to; at each site the user experiences different adventures, shortages, and hardships, and needs to make decisions to survive. The Voyage of the Mimi is based on the story of the 72-foot ship, Mimi, and her crew, who set out on the open seas to locate and study whales. Participants become members of the crew and have to work with them to learn and survive.

Simulation programs develop problem-solving skills as well as specific skills in mathematics and other areas, such as science and geography. For adults, Penn State's Institute for Adult Literacy has developed "A Day in the Life," available through Curriculum Associates, which is designed to build reading, math, and problem-solving skills in five occupational contexts. This highly interactive program exemplifies the use of text, graphics, and animation to put the adult learner in realistic situations in which directions must be read and followed and mathematics understood and used to solve problems in the workplace.

Because of their life experiences, adults bring special skills to simulation situations. Because technology permits a realistic description of the problem through visual and audio modes, there is much potential in this particular use of technology for the teaching of mathematics.

Utilities for Exploration

In this category are programs that have been created to assist with analysis or manipulation of quantitative, statistical, or geometric information, rather than teach a particular subject area. They enable a user to enter data, drawings, or formulas and "work" with them. They usually include a workspace or scratchpad to type in commands, available functions, and provisions for graphing. Through the use of such a package, a user is able to "see" data organized in different ways (e.g., a table, a pie chart, or line graph), execute statistical functions, or apply financial or scientific analysis.

The exhibit areas of any large conference of teachers of mathematics will include displays of many such packages. Some programs available as of this writing include (but are not limited to) Solve (Pacific Crest Software), the Algebraic Proposer (Wm. C. Brown), MathCAD (Mathsoft), and the GeoExplorer (Scott Foresman), which help students explore the areas of geometry and algebra, or the more general Mathematics Toolbox (LOGAL Tangible Math). Also included in this category would be packages that are specifically for statistical analysis, such as Minitab, and business spreadsheet programs, such as Lotus or Excel. Obviously, the applications to which such packages can be put to use are unlimited, and they are equally suited for both K-12 and adult students (although some are meant to be used primarily by more advanced students).

Integrated Applications

Some software packages combine word processing, database, spreadsheet, and graphics and communications functions or modules within the same "integrated" program. Alternatively, software companies sell a "suite" of separate programs that enable easy movements or sharing of data, text, and graphics. Especially with adults, the use of integrated software packages makes sense for those who have had some exposure to software at work (or will be expected to use such software). Integrated software is now available for all computer platforms, is relatively easy to learn, and purportedly will become more user-friendly with each new release.

As mentioned earlier, there has been more integration of writing into mathematics instruction. Students could use the word-processing module of an integrated package to write their learning journal and reports. Databases and spreadsheets are also becoming familiar tools to many adults. Yet, some adults may have used them with templates (or built in "shortcuts") and therefore may not be aware of the full capability of these packages, especially in the use of mathematical and statistical functions. Books and articles are now available that discuss the use of spreadsheets and databases for math instruction. Adult students often become quite engaged in projects in which the data comes from work or is based on a personal interest, like investing in the stock market. The spreadsheet (and sometimes the database) is an ideal tool for analyzing this data, and it enables the development of graphs as well.

Advantages and Disadvantages of Computer Software

Some advantages and disadvantages of the types of software presented have already been discussed. A computer user is an active learner, unless he or she is a passive member of a team. Learner control is high, especially in packages that incorporate the use of hypertext, whereby a user can choose what to work on or examine next. Cooperative learning is certainly an option, although it is up to the instructor to construct a learning environment that is conducive to cooperation.

Does the use of computer software enhance critical thinking? Drill and practice programs often do not, which does raise the question of their overall usefulness. Some games certainly do, especially the ones that are games of logic. Simulation packages hold the highest potential to challenge the critical thinker, and quantitative analysis software can assist the student in answering critical thinking types of questions.

As for added value, this is an important issue for learners. Adults value their time and, for many, additional time and effort will be required to even gain access to a computer. If practice can reasonably be implemented through a workbook, this is the appropriate choice because it is more available. However, the addition of graphics, video, and sound that enhance the understanding of underlying mathematical concepts may convince the adult learner to spend extra time on learning to use a computer.

Of course, the use of any kind of software requires hardware. Special care needs to be taken during the purchase of hardware to check whether it will work with the software planned for use and whether it can accommodate the next wave of computer programs. A current limitation is that the majority of adult students still do not have a computer at home, and it is often difficult to come to a site to use the software regularly. Even if students did have a home computer, they may not be able to use the instructional package available at their adult education program due to copyright regulations restricting use to one machine. Some distributors supply site licenses and may even allow students to bring copies home.

There are creative ways to incorporate limited copies of software and available machines into a curriculum. One way is to have students view the output from one computer on an overhead, using an LCD panel. However, this technique limits the ability to tailor a lesson to a learner's individual learning rate. Students could also be put in teams that rotate during class in use of the computer. Despite the challenges presented by limits on access and use, the promise of modern math software for the adult learner requires that we make every effort to enable learners to use it.

TELECOMMUNICATIONS AND NETWORK RESOURCES

The potential impact of telecommunications on education is just beginning to be realized. Certainly, the addition of a modem to a computer has provided easy access for teachers (but increasingly also for students) to uncountable resources. Electronic mailing lists provide easier access to information about software, books, and other resources. A growing number of electronic discussion groups enable adult educators in distant locations to exchange ideas on a range of issues related to the teaching and learning of mathematics. For example, a Boston-based electronic discussion group called NUMERACY (established by the Adult Numeracy Practitioners Network [ANPN]) now records dozens of messages per week. Online exchanges on the ANPN list, whose members are from the United States, Australia, Canada, Israel, and several European countries, cover an increasingly wide range of topics, such as ideas on teaching the meaning of negative numbers, to attempts to clarify some of the linguistic demands inherent in learning mathematics and ways for linking literacy and mathematics teaching, to methods for explaining the ideas underlying percents or fractions, to options for teaching specific mathematical skills required in the workplace, and more. Members of the NUMERACY list have also recently served as a "virtual" discussion group, joined with local discussion groups of adult educators in different cities in the United States, as part of an ANPN project aimed at developing a consensus over the needed skills and desired curriculum for adult learners of mathematics in the United States.

International computer networks are now allowing access to more and better resources related to the teaching and learning of mathematics and should thus be viewed as an increasingly important area for investment at the local, regional, and national levels. The move to the World Wide Web (WWW), an Internet-based network that transmits graphics (including some video) and sound and which enables publication of fully formatted documents, support materials, and newsletters, has the potential to enable collaborative work of diverse groups of practitioners. In the long term, these networks may remove some of the very walls surrounding our classrooms and enable "distance education" to become a reality for teachers and students alike.

FOUR CASE STUDIES

At this point, the reader may be saying a number of things. Perhaps you feel overwhelmed and have no idea where to start. Perhaps you think that there is no way with your budget to incorporate technology into your teaching. Below are four case studies that may make clearer the alternatives for your next step.

Case 1: The Single Computer

Dear Betty:
    I find all the options you mention fascinating, but at the place where I work, we only have one computer in the room where I teach and I generally have up to twenty people in my class. How can I make that computer useful for such a group?

    Signed, Interested but Doubtful

Dear Interested But Doubtful:
    There are a number of possible solutions to your dilemma. One possibility is to borrow what is called an LCD panel. This is a panel you attach to the computer so that you can project the image from the screen using an overhead projector. In this way, you could at least demonstrate some uses of the computer to the group. Handouts could be prepared to give the student some of the data at their desks. Graphs could then be shown on the overhead and discussed. Of the types discussed in the section on computer software, instructional software (especially drill and practice) and games are probably the least appropriate to use in this manner. Quantitative analysis packages and simulation programs would work quite well, however.
    If an LCD panel is not available, another option is to place your students in learning teams of three or four. Assign each group a multi-step problem for which a computer is needed for a piece of the assignment. Then have groups take turns using the computer. In this way, you can incorporate cooperative learning into your classroom as well as the use of technology. Again, quantitative analysis packages and simulation programs are probably the most appropriate types of programs to use in this format.
    Some packages can be used independently by students, so during out-of-class time, a schedule can be set up for individual or pair work with the computer.
    And, don't forget some other forms of technology, such as calculators, videotapes, and audiotapes. The important goal to consider carefully is how to incorporate technology into your overall educational practice. If you consider the computer just an add-on, your students will as well.

Case 2: Money to Burn

Dear Betty:
 I've just received notice that we are applying for a grant and I need to identify what types of technology we would like to purchase. I've been told to aim for a total of $5000.

 Signed, Gotta Spend It

Dear Gotta Spend It:
 That's great! I hope you get the grant!
 The key to creating the appropriate package, no matter what the funds are, is to identify what you wish to do. It is crucial that you not be too near-sighted, however. Think at least three years ahead. Once you identify your educational goals, then you are ready to select your technological tools. For example, if you have primarily students preparing for business training, your educational goals will differ from a program that has primarily students aiming to improve their basic skills and obtain a high school equivalency diploma. At least one computer is a likely item, but the type will depend greatly on what kind of software you wish to use. The Macintosh is still the best computer for graphics-oriented programs. Yet, with Windows for the IBM-compatible and now the new system that combines IBM and MAC, this decision may be much easier in the future. You should seriously consider a CD-ROM drive in your computer system. In fact, the capability and availability of CD-ROMs warrant selection of computers that include built-in CD-ROM drives. You also may want to seriously consider acquiring a LCD panel, as mentioned to Interested But Doubtful.
 For software, be sure to establish criteria for selection before deciding what to buy. You may want to consider the criteria identified earlier in addition to your own criteria. It is very important that your educational goals hold the highest priority, with the technology viewed with respect to how it may best serve those goals. Consult your literacy resource center or connect with your local school system to check on recommended software or suppliers.
 Although computer hardware and software are common components, do not forget to consider other possibilities, such as calculators and videotapes. As mentioned in this chapter, these can be very powerful tools.

Case 3: Can I Really Do This?

Dear Betty:

I found many suggestions in your chapter exciting to consider, but I must confess that I'm one of those people who feels intimidated by technology. Is it really possible for me to effectively use technology with my students?

Signed, A Techniphobe

Dear Techniphobe:

First of all, let me assure you that you are not alone. You will find among your students a full range of attitudes and abilities connected with technology. In fact, you may want to begin by polling your students about how they feel about technology.

The best strategy for you would be to begin slowly. A wide range of technologies have been mentioned in this chapter. How about starting with a taperecorder? As mentioned, there are a variety of projects you can do with your students in this area. In using the VCR, there is no need to begin with programming. Many useful videos can be viewed with the play button. Additionally, there are a full range of calculators to use, and you may want to consider attending some calculator workshops. As your comfort level increases, so will your ability to use a wider range of technological tools. Perhaps you could pair with another teacher and together you can experiment with a simple software package. Your students will be inspired by your willingness to experiment and learn from your mistakes.

One important criteria for evaluating computer software is user friendliness. How easily does it teach you an your students how to use it?

Case 4: Computers and Computers

Dear Betty:

I've recently been able to arrange for our adult learners to have access to the computer lab at a local college. The room has 10 computers, all networked, with individual CD-ROM and floppy drives and a shared printer. I have about 30 students. What now?

Signed, Excited, but anxious

Dear Excited but Anxious:

Congratulations! It sounds like you have negotiated a very nice opportunity for your students.

An important aspect of any computer lab is how the equipment is arranged in the room. If at all possible, avoid a situation in which the computers are in rows such that it is difficult for you to maneuver through the room to see how students are doing. If the room is this way, see if you can convince the college to consider the advantages of a change. I strongly recommend having the computers along the perimeter of the room. In this way, you can have the students form a circle facing you to discuss concepts and also to allow for easy use of the computers by teams. From the center, you have a good view of and access to every team.

What types of software should you use? In your case, this decision will be greatly influenced by what is available. Be sure to use the criteria given earlier in making your decisions as to what to use. Software that enhances group problem solving is a good candidate. I would not recommend drill-and-practice beyond a brief activity to assess the math skills of the individual teams. The more carefully you consider your goals and develop activities that meet these goals, the more effective will be the use of these valuable tools.

Good luck and have fun!

IMPLICATIONS

In this chapter I reviewed some of the reasons to incorporate technology into programs for adults learning mathematics and reflected on the advantages and challenges connected with implementing technological tools in the mathematics classroom. The need to incorporate technology into the mathematics education of adult learners is clear. Through the use of technology, lifelong learning can be more easily achieved. Technological tools enable a wide range of adult learners with diverse learning needs to improve their skills and understanding. Technology encourages active learning and can contribute to the development of critical thinking skills that have a greater chance to be transferred by learners to new situations. Technology can provide the opportunity to experience a "virtual field trip" and to examine functional math applications, including those used in the workplace.

The use of technology can increase the learner's control of the learning environment, and flexibly combine group-based as well as independent learning experiences. In this context, teachers will increasingly become facilitators for learning, as opposed to the source of knowledge, and will need to consider how to foster an environment that encourages learner self-assessment. Because self-assessment is essential for the lifelong learner, yet unfamiliar to adult learners who have studied math before in a traditional school environment, the development of self-assessment skills (and tools) will be a challenging yet worthy goal.

One of the major challenges that comes with educational technology is that it changes so quickly. It is unlikely that the resources a teacher purchases today will be on the cutting edge for long, if they ever will. However, if they meet the goals that were established for the program, they will have served their purpose well. That said, the future for uses of technology in adult education is exciting. With the increased capacity of computers and the growth of faster and larger mass-storage devices, better integration of full video, sound, text, and graphics will be possible. These possibilities, combined with a better understanding of how adults learn and how they learn mathematics, should enable the creation of more flexible, realistic, and interactive computer programs for adults learning mathematics. Imagine being able to make a mistake on a simulated job without getting fired! The improved capability of hand-held devices will increase portability, which is so important to adult learners. With fiber optic networks bringing increased delivery of instructional video programs (perhaps one day interactive) to the home, opportunities for learning mathematics may become even more available.

Even though burgeoning computer and network applications offer great promise, the use of simple or "old" technologies has a lot to offer and may be very cost-effective in specific teaching situations. The real challenge will be to focus on the curriculum goals, rather than on the technology that delivers the curriculum. As exciting as the technology may be, it must remain a means to the end, rather than an end in itself.

REFERENCES

Buerk, D. (1985). The voices of women making meaning in mathematics. *Journal of Education, 167*(3), 59-70.

Connolly, P., & Vilardi, T. (Eds.) (1989). *Writing to learn mathematics and science*. New York: Teachers College Press.

Gal, I., & Schuh, A. (1994). *Who counts in adult literacy programs: A national survey of numeracy education* (Tech. Rep. No. TR94-09). Philadelphia: University of Pennsylvania, National Center on Adult Literacy.

Grandgenett, N. (1991). Roles of computer technology in the mathematics education of the gifted. *Gifted Child Today, 14*(1), 18-23.

Hembree, R., & Dessart, D. (1992). Research on calculators in mathematics education. In J. Fey & C. Hirsch (Eds.), *NCTM Yearbook* (pp. 23-32). Reston, VA: National Council of Teachers of Mathematics.

Johnson, J., & Smith, K. (1993). *Cooperative learning.* Minneapolis, MN: Cooperative Learning Institute.

Levesque, J. A. (1989). Using computers to motivate learners. *Social Studies and the Young Learner, 2*(1), 9-11.

National Center on Adult Literacy (NCAL). (1994). *Voices from the field: The use of computer technology in adult literacy* (pp. 1, 7-8). Philadelphia, PA: Author.

National Council of Teachers of Mathematics (NCTM). (1989). *Curriculum and evaluation standards for school mathematics.* Reston, VA: Author.

Office of Technology Assessment. (1993). *Technology offers new ways to boost literacy.* Washington, DC: Author.

The Secretary's Commission on Achieving Necessary Skills (SCANS). (1991). *What work requires of schools: A SCANS report for America 2000.* Washington, DC: U.S. Department of Labor.

Steinberg, E. R. (1991). *Computer-assisted instruction: A synthesis of theory, practice, and technology.* Hillsdale, NJ: Erlbaum.

Stenmark, J. K., Thompson, V., & Cossey, R. (1986). *Family math.* Berkeley: University of California, Lawrence Hall of Science.

Turner, T. C. (1993). *Literacy and machines: An overview of the use of technology in adult literacy programs* (Tech. Rep. No. TR93-03). Philadelphia: University of Pennsylvania, National Center on Adult Literacy.

9

Teaching Mathematics to Adults with Specific Learning Difficulties

Martha Sacks
Dorothy M. Cebula
Temple University

Adults with histories of learning difficulties present a complex series of teaching challenges to the Adult Basic Education instructor. This chapter presents techniques for working collaboratively with adults who have had the experience of extensive instruction but are unable to consistently succeed in math. It addresses assessment and instructional approaches aimed at overcoming reading, visual and auditory, perceptual, language processing, conceptual, and organizational concerns.

INTRODUCTION

The teaching of adults with histories of learning difficulties provides one of the greatest challenges to the ABE mathematics teacher. Unlike similarly instructed classmates, they are unable to consistently demonstrate mastery in mathematics or related areas. In developing effective instructional approaches, a prospective teacher needs to recognize the learner's mindset and the effects of potential perceptual, linguistic, and processing barriers. Many learners have developed fears and anxieties related to math that are extraneous to direct problem-solving processes. The teacher should be sensitive to the many aspects of the learner's attempt to understand math strategies, concepts, and ideas.

Although some ABE students may have been tested and identified as "learning disabled," there may be many students who have learning differences that were never recognized or diagnosed. The term "learning disability" includes a large, heterogeneous group of perceptual, organizational, conceptual, and linguistic conditions that affect understanding, processing, or retrieval of information. Since the first use of the term in 1962, the definition of learning disabilities has been the focus of controversy in the field of education. "Learning disabilities" has emerged as an overinclusive field that provides little clarification for the specific nature of problem areas. This chapter uses the term "specific learning difficulties" in place of the term "learning disabilities" to describe particular barriers to learning and areas that remain problematic to a learner to a severe degree beyond what could usually be anticipated for similarly instructed, motivated peers of average cognitive aptitude.

Part II of this chapter describes approaches to assessment and identification of math problem-solving difficulties and highlights the need for a multifaceted approach to assessment aimed toward differentiating specific learning difficulties. Part III provides more details about six specific types of learning difficulties: reading, visual perception, auditory processing, language processing, cognitive and conceptual, and organization and time management issues. For each of these separate but related types, techniques are outlined that have proven helpful in making math learning goals achievable, even for adult students with specific learning difficulties.

ASSESSMENT

Adult educators often ask, "How can I assess specific learning difficulties in the area of mathematics?" Formal diagnostic tests are often used by adult education programs to inform placement decisions, but when specific learning difficulties are suspected, additional assessment strategies are called for. This section introduces the techniques of error analysis, talking aloud or protocol analysis activities, and clinical interviews, which are often necessary to effectively identify individual problem solving strategies, learning styles, and performance areas in need of development.

Error Analysis

Standardized tests provide information about levels of mathematical achievement and types of procedures known to the student, but usually do not provide information about the causes and specific nature of problem solving difficulties. After scoring a test, an analysis of errors on test items can enable the teacher to identify key types and patterns of mistakes and allow the instructor to decide on spheres for direct instruction. Specifically, mistakes in calculations will require a different instructional focus than would errors resulting from distorted alignments of numerals, misread signs or symbols, or examples left unanswered. However, problem-solving strategies cannot be determined with this method alone because there is usually no direct communication with the student, and thus no description of the rationale for responses.

Talk-Aloud Activities

To gain insight into problem-solving strategies, students should be encouraged to explain their thinking process while solving a problem. With some training and practice, a teacher can glean meaningful information from statements made as the learner interacts with and verbally explains each step of the problem. An extension of this approach is protocol analysis (see Parmar, 1992), which includes a dialogue and a series of questions that help students articulate reasons behind the approach they chose for solving a given task. The analysis of data from this structured conversation can help identify several areas of potential processing difficulties, such as reading proficiency, identification of needed operations, use of manipulatives, knowledge of basic computations, or facility to self-correct, concentrate and persist; all of these can impede the ability to think quantitatively and logically.

Clinical Interview

The use of formal testing, analysis of specific error patterns, and use of verbal protocols can suggest a tentative profile of strengths and weaknesses that affect a student's problem-solving and reasoning processes. The clinical interview is needed not to provide additional information about specific problem-solving issues but rather to portray the student's general perspectives on (mathematics) learning. An effective interview will help describe key issues such as prior experiences with mathematics learning, strategies attempted in past math learning and assessment contexts and whether or not they worked, applications of mathematics to daily living, the student's academic goals, expectations about the current learning situation and motivations propelling current participation in an adult learning episode, and the learners' attitudes and feelings regarding mathematics learning.

For example, consider this list of questions.

- How would you describe your past experiences with learning mathematics?
- What aspect of mathematics has given you the greatest difficulty?
- What approaches have you tried in learning math (or other subjects)? What worked best for you? What approaches did you dislike? Why?
- How do you solve math problems?
- How do you approach questions involving math?
- Can you suggest ways that would help you in learning or doing mathematics?
- When people talk about learning mathematics, what do you think of?
- Why are you interested in working on mathematics now? Are there particular areas of math that are more important to you in your job or family responsibilities?
- How do you plan to study and practice math between classes?

We recommend a combination of the three assessment approaches just discussed to most effectively evaluate the existence of specific learning difficulties and chart a course for instruction. (To be most helpful, the interviews and other student responses should be tape-recorded and/or transcribed.) The results of the evaluation can be described using categories such as goals and needs, motivations and expectations, perceptions about math, strategies and approaches to problem solving, and possible learning difficulties. Instructors who can implement this multifaceted approach to assessment, as well as listen and learn from adults about their experiences, will be able to help empower learners to develop effective learning strategies.

"How can I distinguish specific learning difficulties from other problems often seen in adult basic education classes?" Many ABE students cite inadequate practice, limited motivation, or poor instruction as reasons for poor prior success in mathematics. However, persons with histories of specific learning difficulties (including perceptual, organizational, and conceptual issues such as those discussed later) more often report other or additional concerns. They may mention, for instance, problems related to memory and the ability to consistently recall or apply number facts, mystification with the language and reasoning processes used in learning math, or repetitive instruction and extensive hours of meaningless labor. Many may report problems relative to blurred or distorted vision, although acuity is considered adequate. Some may have difficulty reading (decoding or comprehending) mathematical texts or materials. In testing situations, many report lack of time to finish, errors in copying problems or answers, inaccurate computing, or difficulty in memorizing formulas. Often these reports belie inherent difficulty with intuitive understanding of concepts involved and may ignore or hide underlying difficulty with abstraction and problem solving.

The information provided by the students during the interview will assist the instructor in differentiating between individuals with specific learning difficulties and other ABE students who have had difficulty in mathematics in the past. By observing students in testing situations and listening to their stories of math experiences, an ABE teacher will be better equipped to distinguish specific barriers to learning from problems related to limited prior instruction. Adults with characteristic problems of specific learning difficulties should not be expected to fit a standard mold. Many will have combinations of the difficulties noted while some who have made extensive efforts at compensation may have few difficulties, although the need for individual adjustments in teaching may be as crucial.

INSTRUCTION

Given the pressures of time and multiple goals, and concerns about the ability of students with specific learning difficulties to acquire, maintain, and apply mathematics skills, the ABE teacher is faced with a formidable, yet achievable task. Once a baseline for starting instruction is determined through interviews and assessment, the teacher and learner should jointly create a plan for developing mathematical skills with the goals of the learner and the program in mind. If gaps appear in the learner's skill profile (identified through diagnostic testing), it will be necessary to develop competency in these areas as well. In any case, sensitivity to the unique nature of the student's learning needs must be maintained. For example, if a student has indicated severe problems with memory (see later) that have restricted recall of number or multiplication facts, practice in effectively using a calculator may be necessary before returning to an initial goal of developing problem-solving strategies. The most efficient method is to work from the individual's strengths and help each student find ways to circumvent the weaknesses.

This section first considers some general issues involved in developing an action plan for students with specific learning difficulties and then focuses on six specific areas: reading, visual perception, auditory and language processing, cognitive and conceptual, and organization and time management issues. Each area is briefly explained, and specific techniques or instructional adjustments are suggested.

General Issues

Many instructional approaches helpful for teaching students with specific learning difficulties are also helpful to all ABE learners, and vice versa. Suggestions for working with ABE students are included in this volume, as well as in many instructors' guides that accompany standard commercial texts or "basal series" used in adult education (and K-12) instructional programs. Relying solely on commercially published materials, however, can lead to student and instructor frustration because individual instructional objectives are not easily separated from the broader sequence of lessons in traditional (basal) textbook series. This issue was raised in a study by Engelmann, Carnine, and Steely (1991) that summarized research on math basals and identified inadequacies that might contribute to the poor performance of students. These inadequacies included:

- Marginal provisions to ensure knowledge of prerequisites
- Rapid rate of concept introduction
- Inconsistent presentation of problem solving techniques
- Unclear or confusing explanations
- Inadequate transitions from guided to independent practice exercises
- Limited opportunity for practice or review

Because most math textbook series are similar in design, these shortcomings can be considered the norm. Students with specific learning difficulties could find exclusive dependence on basal texts inadequate to meet their instructional needs.

In contrast to a standardized program or basal series with little flexibility and limited options for review, the vision of mathematics teaching advocated by the National Council of Teachers of Mathematics (see chap. 3, this volume) supports cooperative learning with students working together to conjecture, invent, and problem solve with the goal of developing mathematical reasoning and communication skills. The learning environment recommended by NCTM would benefit all students, including students with specific learning difficulties.

Although classroom experiences that embody the vision advocated by NCTM can serve a wide range of learners, adults with histories of failure in mathematics need to have their specific concerns addressed directly. Instructional adjustments designed to address specific learning difficulties are described further below, but there are also general guidelines that can be implemented for and assist practically all learners who face perceptual, organizational, and conceptual challenges:

- Begin each class with a review of the previous material.
- Start each new topic with a review of any past material that is necessary for the new topic.
- Provide an outline of each lesson and highlight the relevance of the lesson to the life of the learner.
- Solve practice problems together with students; clarify patterns similar to several problems.
- Present material in a clear, sequential manner; relations should be specifically explained.
- Instruct in and encourage effective use of calculators.
- Suggest the use of taperecorders for specific parts of lessons. Be certain that supplemental visual materials are also available.
- Make available a legible, complete copy of notes for each lesson.
- Emphasize key points during and summarize key points at the end of each lesson.
- Before a testing situation, review item samples that are comparable in format, context, and content to any actual exam items to be administered.
- Provide extended time or remove time constraints on tests or class tasks, as needed, so that a test is not a measure of speed but of understanding.

Awareness of each student's unique learning needs is essential for the development of individual educational strategies. Next we focus on six specific categories of perceptual or processing difficulties. Although some individuals may have characteristics that overlap several of the categories, few will demonstrate all of the difficulties noted.

Reading Difficulties Influencing Math Instruction

People who have difficulty with reading comprehension or decoding frequently resist or avoid math activities in which text is required. They often misread problems or struggle over words instead of focusing on the meaning of an entire phrase or sentence. When assessing math performance, the teacher who can recognize the manner in which a student reads will have a key to unlocking a major stumbling block to learning. If a student sounds each word laboriously or states concerns about additional frustrations with reading comprehension, an instructor should be prepared to evaluate the influence of reading on the math task (see O'Mara, 1981, for further discussion of such issues). Students who can explain the process used in a problem when the question is presented orally may not require extensive instruction in a particular math task. Their time could be better spent in developing reading strategies.

- If auditory reinforcement is helpful, provide a taped text to facilitate the student's understanding of word problems.
- Direct instruction with carefully planned step-by-step explanations of problems and terminology can facilitate understanding.

Visual Perceptual Difficulties

Many people with histories of school-related difficulties describe problems involving visual processing that include recognizing, conceptualizing, or remembering visual patterns. This could affect identification or recall of mathematical symbols or formulas. Digit sequence distortions can also occur, or the numbers can be reversed, transposed, omitted, or repeated. Some people report difficulty with remembering symbols when copying information; therefore, they develop an extensive digit-by-digit transfer method. Frequently, copying and aligning columns of numbers becomes a monumental task. Occasionally, learners will begin a series of operations involving addition, for example, and then miss a shift in operations signaled by another symbol such as a multiplication sign. In many situations, these perceptually based markers can be identified by analyzing errors on regular tests (or teacher-made problems). Most adult learners who are aware of visually based errors can vividly describe their frustrating experiences, but continue to make such errors unless reminded of the need to review material carefully.

Copying.

- Rather than using extensive time to transfer examples from a text, prepare copies of examples in a clear, well-organized format to focus efforts on problem solving itself. At a later time, prudent copying of a limited number of examples may be encouraged if practice appears beneficial.
- Provide a set of notes for each lesson so the student will be able to concentrate on the activity rather than the process of notetaking. This can reduce frustration and enhance the opportunity for learning by reducing writing pressures.
- When transposing or other errors with copying digits are frequent, encourage the learner to orally read numbers and match with the intended original. This should be done as a separate activity from the computational exercise. Use a large colored index card to block other print to focus on digit series.
- Encourage the use of graph paper to align columns of numbers. If graph paper is not readily available, turn ruled paper on the side to provide lined support.

Symbol identification.

- Use talking calculators (or a talking device that can read exercises with bar-codes under them) to check examples. The vocalization is helpful in allowing a learner to discriminate differences in executed versus intended number patterns. If this type of calculator is not available, tape record planned examples to check the accuracy of copied material.
- Use highlighters or colored chalk to emphasize examples or key aspects of a problem. Some learners find that routine highlighting of operation signs helps to direct them; the use of different colors for basic operations enhances discrimination efforts.

Test adjustments.

- If standardized testing is required, offer extended time. In most situations, approximately double the usual time limit should be given. This will enable most students with perceptual problems to work through each problem carefully and relieve the anxiety often produced by time constraints.

- Provide graphing paper for those who have difficulty aligning numbers.
- Use enlarged test pages so the problems do not have to be copied onto a separate sheet of paper.
- Encourage use of a marker to assist in tracking.

Auditory Processing Difficulties

Many students have difficulty comprehending auditorily presented information. This involves memory and language processing. Problems may include attempting to follow a verbal explanation of a problem and forgetting the steps before writing notes. It may also involve distorting the sequence of steps by following directions out of order. Difficulty blending sounds together may result in poor reading performance which may make a math text indecipherable.

Notetaking.

- Distribute an outline for each lesson so the student will be able to follow the lesson visually as well as auditorily.
- Have the learner take notes in a controlled setting. Review the notes collaboratively to ensure that they contain complete information.
- Tape record lesson activities. Have the student control audiotape recordings so that only specific lesson-related information is recorded. Encourage repetition of the tape and essential transcription if necessary.

Test adjustments.

- Evaluate the need for extended time if standardized testing is required. This will give students with reading difficulties time to complete the reading portion of each exam.
- Read word problems aloud to augment independent decoding if reading errors would affect the understanding of the process.
- Repeat directions carefully and indicate taped directions that can be used as a reference.

Language processing difficulties.

Mastering the language of mathematics may be a struggle, yet it is essential to learning mathematics (Laborde, 1990). Expressive language difficulties may be noticeable when listening to explanations of problem solutions if a person is unable to explain or describe examples and procedures. Other linguistic concerns may be evident in recognizing multiple meanings of words or comprehending specific use of terms. Confusion may also arise when taking notes if individual words and ideas require extensive processing; subsequent information can also be missed.

Another complication involves interpretations of subtleties of word problems. A person may be competent in decoding print and oral presentations but confuse the "language messages." Understanding language use in word problems is essential for effective problem solving. Because this requires reading and/or listening, an instructor should look for additional clues to gaps in language facility. The student who cannot recognize terms that trigger an operation may need direction in "translating" phrases, as in word problems, into mathematical procedures. For example, "the sum of these numbers is . . ." should construct the image of a series of numbers followed by the total value. Some adults also need explanations for more concrete terms that are used almost exclusively in math, such as "denominator" and "multiplicand." A student's limited knowledge of the definition and application of such words can usually be more easily addressed with direct instruction.

It is helpful to listen to the way a person speaks while explaining information. A student who tends to speak in short, halting phrases may be signaling a preference for brief conversational segments in return. A person who speaks extensively without providing much information may have difficulty with precise language and lack the vocabulary necessary for effective math learning. The learner who is unable to accurately retrieve terms but attempts to explain by referencing other ideas may need specific attention toward developing the language of mathematics.

Language terminology and use.

- Present key terms in writing as well as in conversation. With direct instruction, explain the term and expected occurrences. Explicitly teach every step in the translation process and provide detailed correction procedures. Have the student identify key terms in sample problems before attempting to solve problems. The student may also benefit from writing examples with step-by-step solutions including appropriate terminology.
- Encourage the learner to highlight or circle key terms that suggest operations in a problem. Using visual supports, such as a

copy of a problem on a board or large card, may help reinforce the use of each word as a key element in the structure of a math sentence.

- Present written problems written in predictable styles. Introduce different patterns only after the student has developed a comfortable grasp of the terms and expectations of a single pattern.

Memory.

- Encourage the student who forgets specific terms to prepare cards with necessary terms, definitions, and examples. As new terms become necessary, the card file can expand. As terms become more familiar, they can be moved to a secondary section. This system can be helpful for encouraging practice and review when limited time is available. It allows handy reference when a student needs assistance.

Processing.

- When language demands appear to overload a student's comprehension, reduce the length of sentences being presented. This could involve rewriting math problems or eliminating unnecessary phrases. In discussion, reduce verbal explanations to short sentences spaced with pauses to help understanding. In situations in which language processing is compounded with new learning, the teacher should be careful to allow extensive time for thinking. Providing visual examples with oral explanations can also help offset frustrations of verbal overload.
- Prepare the learner for the patterns involved in explanations. For example, before attempting to explain the process involved in a problem, tell the student what steps will be taken to teach a lesson. Outlining general steps will help guide learning and eliminate the need for guessing.
- Watch for facial cues. If a student cannot understand verbal information, facial expressions can alert an instructor. Teachers should be able to recognize when students require rewording or alternate explanations based on nonverbal behaviors.

Test adjustments.

- Encourage the use of a card with necessary formulas.

Cognitive and Conceptual Difficulties

Cognitive and conceptual difficulties are often reflected in problems with tasks involving abstract reasoning. Although learners may repeat the language used to describe a concept, they may be unable to consistently distinguish appropriate situations when rules or concepts belong in or apply to a given problem, and they may have difficulty organizing approaches to a new problem. A person with such a profile usually has much difficulty mastering the intent and essence of mathematics.

Frequently, a student with reasoning difficulty will describe methods of study that depend heavily on rote memory. The learner may devote hours to completing several problems of the same type yet be unable to transfer that practice to a differently presented problem. One can speculate that the student has not internalized the process or rationale involved in following a series of procedures to solve math problems and does not have an intuitive grasp of number sense. In other words, a person may "plug in the numbers" in an example but may not be able to explain a reason for choosing certain numbers or using certain operations.

A related problem involves the difficulty in grasping concepts related to numbers. When a person cannot differentiate between "one hundred" and "one thousand," the effectiveness of estimation as a self-testing device is limited. Strategies to address the difficulty in distinguishing concepts require attention before extensive assistance in problem solving will be effective.

An additional difficulty can be seen in the problem of mentally shifting from one concept or idea to another. If a student is performing tasks that involve addition, for example, it may be difficult to shift to a mode in which subtraction problems are being presented. (This may also be a factor when students have difficulty understanding the relationships between operations, for instance, understanding multiplication as repeated addition.) The instructor who can identify the nature of the difficulty can distinguish between assumed errors in computation due to carelessness, perceptual distortions, or counting mistakes caused by an inability to shift mental conceptualizations.

Abstract reasoning.

- Present material in a clear, sequential manner, specifically explaining relationships between steps in a problem.

- Do not skip steps when explaining problems.
- Review previous material before moving to new operations, but be careful to highlight differences.
- Individualized assistance may be necessary to aid the student who has difficulty developing abstract concepts. It is often helpful to encourage discussion of a word problem or the steps in an example. When relating two comparable problems, an instructor may be able to explain similarities without using specific numbers. The use of abstract language may help the student move from concrete to more global thinking patterns.
- Utilize the "concrete-semi-concrete-abstract" (CSA) teaching sequence, described in detail by Miller, Mercer, and Dillon (1992). The CSA sequence includes instruction in concrete (use of manipulative objects) to semiconcrete (use of graphic representations with pictures of tallies) to abstract (use of numbers only). This type of sequence increases the student's understanding of concepts.
- Instruct students in heuristic strategies to increase conceptualization of math problems. Heuristic strategies, such as the transcription of the problem into a different format, trial-and-error responses, or the development of analogous problems, are problem-solving strategies that enable students to approach word problems with the understanding that there are alternative solutions and alternative routes to the problem (refer to Giordano, 1992, for further information).
- When addressing word problems, encourage the student to identify patterns in the question rather than focus on individual words that may vary from problem to problem. Try to encourage verbal or written comparisons between similar problems to stimulate and reinforce abstract thinking.

Conceptual focus.

- Begin each session with a clearly articulated review of previous material. Explain the general principle involved and relate that information to the upcoming activity. By addressing topics according to a more general rule, the student who needed assistance in thinking abstractly will be encouraged to link information more quickly.
- Explain relations between ideas or problems in a structured and sequential manner. Indirect, discovery learning methods are frequently unsuccessful with students who have difficulty developing concepts, and students fail to develop transfer learning.

- Do not skip steps or suggest "short cuts" when demonstrating problems. The student who experiences difficulty with abstract concepts may be easily confused if steps are not identified or if changes are made. When modifications are appropriate, explain the process involved in making changes and demonstrate how two different methods can conclude with the same results.
- Continued review may be necessary to enable the student to remember previous material and progress to new material. In many situations, a student who has difficulty with the concepts of a new idea may require repeated practice in solving problems before mastery can be expected.
- Self-instructional strategy training can provide students with a set of verbal prompts designed to assist them in problem solving. Montague and Bos (1986) described one method of cognitive strategy training that includes the following verbal math problem-solving strategy: read the problem aloud, paraphrase the problem aloud, visualize, state the problem, hypothesize, estimate; calculate, self-check. Have the student list these steps on a card or book cover; each step should be checked before moving to the next step.

Organizational and Time Management Difficulties

The student who reports extensive difficulty finding a time and place to study may have a serious barrier to effective success in math learning. When students repeatedly express problems organizing and planning time, assistance must be provided before intensive math instruction can be effective. The student who rarely maintains appointments or cannot locate papers may be in need of direction and suggestions in order to benefit from ABE support.

To assist students in developing time-management strategies, encourage a preplanned routine for study activities. If a student can identify situations that may work best for quiet study, effective practice can be developed with reinforcement from the instructor. Encourage planned study sessions of 15 to 30 minutes to ensure an initial sense of achievement. Provide students with a chart that they can use to organize their weekly schedule.

Organization of materials can be a challenge to math learners. Encourage learners to use specific types of paper rather than random scraps. Spiral notebooks are useful in maintaining order. Note cards, colored if necessary, may be suggested if the student also needs assistance with practice activities.

IMPLICATIONS

Adult students with a history of learning difficulties may need assistance in developing strategies for mathematical problem solving and applying meaning to mathematical information, in or out of school. The application of strategies and techniques presented in this chapter can assist in developing greater sensitivity to the specific challenges encountered by learners with histories of learning difficulties. The teacher and learner need to work as a team to be better able to recognize specific barriers and develop approaches to compensate or correct misconceptions and errors.

Past difficulties do not necessarily eliminate the opportunity to achieve the learner's mathematics goals or an ABE program's objectives in the area of mathematics. Yet, the teacher needs to make sure to become familiar with the learner's prior educational history and to recognize the learner's habitual approaches to new learning situations, as well as any negative attitudes or beliefs about self and the domain of mathematics. The teacher and the learner who has experienced learning difficulties in the past should jointly construct a solution-oriented approach to learning and identify strategies to facilitate planning, executing, and monitoring learning and problem-solving processes that take into account perceptual, linguistic, and processing factors that may exist.

REFERENCES

Engelmann, S., Carnine, D., & Steely, D. (1991). Making connections in mathematics. *Journal of Learning Disabilities, 24*, 292-303.

Giordano, G. (1992). Heuristic strategies: An aid for solving verbal mathematical problems. *Intervention in School and Clinic, 28*, 88-96.

Laborde, C. (1990). Language and mathematics. In P. Nesher & J. Kilpatrick (Eds.), *Mathematics and cognition.* New York: Cambridge University Press.

Miller, S. P., Mercer, C.D., & Dillon, A. S. (1992). CSA: Acquiring and retaining math skills. *Intervention in School and Clinic, 28*, 105-110.

Montague, M., & Bos, C. (1986). The effect of cognitive strategy training on verbal math problem solving performance of learning disabled adolescents. *Journal of Learning Disabilities, 19*, 26-33.

O'Mara, D. A. (1981). The process of reading mathematics. *Journal of Reading, 25*(7), 22-30.

Parmar, R. S. (1992). Protocol analysis of strategies used by students with mild disabilities when solving arithmetic word problems. *Diagnostique, 17,* 227-243.

10

Writing About Life: Creating Original Math Projects with Adults

Karen Hicks McCormick
Elizabeth Wadlington
Southern Louisiana State University

This chapter presents a model for integrating learning of writing, reading, speaking, and listening with learning of mathematical concepts in ways that are meaningful for adult students. Three questions guide this chapter: How can adult educators integrate mathematics and language arts skills so that students perceive learning as a whole rather than in distinct, isolated parts? How can adult educators make mathematics relevant to students' daily lives so they become confident, competent problem solvers? How can adult educators provide activities that teach language processes and mechanics in such a way that learning is transferred to other areas, including mathematics and real life?

INTRODUCTION

This chapter demonstrates that integration of mathematics with language arts can support both disciplines in a purposeful manner that makes learning relevant to students' lives and to the real world. The strategies used to achieve this can be applied to all types of adult basic skills classes. The chapter is organized in three parts. Part I provides a rationale for the integration of instruction in mathematics and language arts. Part II illustrates such integration in an applied setting, using concrete examples from work with an adult graduate equivalency diploma (GED) class that used a process writing approach, and presents sample lesson plans. (Some readers may prefer to first read Part II before returning to the theoretical background in Part I.) Part III elaborates various methods to embed writing and communicating activities in an adult mathematics classroom.

RATIONALE FOR INTEGRATION OF LANGUAGE ARTS AND MATHEMATICS

The integration of mathematics, reading, and writing in instructional assignments that are pulled from the students' lives appears paramount as it can lead to effective, successful, and functional learning. Support for this argument can be found in literature pertaining to learning theories and specifically to the cognitive and affective domains of learning, standards set by educational professional organizations, economics of the work force, and educational pragmatics. These viewpoints are discussed below.

Support From Learning Theories

The constructivist learning theory (see chap. 2) states that students learn best when they are involved and empowered to interact with the assignment, blending their past experiences with the new information. Weaver (1988, p. 26) uses the term "socio-psycholinguistic model of learning" for the same concept, which stresses the process of reading as a transaction between the reader-listener and the written-spoken words. Every student brings prior knowledge, or schema, to each learning experience that affects understanding. Indeed, the learning is actually constructed by the student when previous learning experiences and the new words intertwine. Tapping into this schema enriches the educational experience.

Holdaway (1986), a New Zealand educator, has identified four elements necessary for literacy learning to take place within the constructivist theory of learning framework. These elements include observation or demonstration, participation, role playing or practice, and performance.

Cambourne's *Seven Natural Learning Conditions* (1988) strongly indicate that students learn best through immersion in the discipline, demonstration of skills, responsibility, student employment, acceptance of approximations, high expectations by teacher and students, and immediate feedback. The constructivist theory of learning is embedded firmly within these learning conditions as the students are responsible and are actively doing the work, accepting the consequences, and receiving feedback to improve. Using process writing in the teaching of mathematics, which is the focus of Part II, demonstrates all of the elements in both Holdaway's and Cambourne's models for effective teaching.

Support From the Cognitive Research Domain

A common concern of students is the irrelevance of their course work in their daily lives. Much of their instruction is based on textbooks used in isolation from real living (Jacobs, 1989). Schools are criticized because students learn problem-solving techniques very mechanically with little transfer ability. If students recognized the connections between mathematics learned in and out of school, their understandings in both settings would be better (Hiebert & Carpenter, 1992). An interdisciplinary curriculum is needed to make learning more relevant, less fragmented, and more stimulating to students.

Poor reading skills are the first roadblock to overcome for successful mathematical problem completion. Regardless of computational skills and mathematical conceptual understanding, if written problems cannot be read, the student cannot go through the steps of working problems. As Cunningham and Ballew (1983) suggest, successful completion of written mathematical problems requires independent reading and interpretation skills, in addition to computational skills. However, mastery of each component skill does not ensure that learners can integrate all skills so as to manage the whole task.

The integration of writing and mathematics has also been shown to predict success with mathematics in school. Silverman, Winograd, and Strohauer (1992) and Winograd (1993) found that "high-status students" in school mathematics clearly define questions that guide problem composition. They also self-monitor their comprehension of problem-solving tasks, and when comprehension falters, they use strategies in a systematic manner to improve their understanding of the problem. In contrast, "low-status students" do not use focus questions to guide writing and write rather ambiguously.

The National Writing Project (University of California-Berkeley) encourages "writing across the curriculum" to improve thinking skills in academic subjects. This instructional reform originated with the belief that the kind of writing students do in school has a direct influence on the quality of thinking in which they are required to engage (Applebee, 1981; Newkirk & Atwell, 1982).

Other research examined how specific types of reading and writing influence students' reasoning, as well as the increased understanding that comes with composing in conjunction with reading (Flower, 1987; Tierney, Soter, O'Flahavan, & McGinley, 1989). A circular pattern in which reading and writing are highly supportive of one another emerges when writers become readers as they reread their text, peer edit by reading other students' writing, and respond to suggestions by writing again (Calkins, 1986).

Support From the Affective Domain

Peer conferencing, freedom to choose topics, a risk-free environment, and divergent assignments are key elements of process writing. These elements motivate, build ownership, encourage open-mindedness, and validate students' life experiences. A sense of community is developed, and attitudes improve as students cooperate to share and respond in groups (Atwell, 1987; Calkins, 1986, 1994; Goldberg, 1986; Harste, Short, & Burke, 1988; Routman, 1988). Kohn (1993) states that the key to changing student apathy into active engagement may be as elementary as permitting students to make decisions about their learning. Kohn goes further to state that the best predictor of workplace burnout is powerlessness—lack of control over what one is doing. Process writing empowers students and heads off burnout; consequently, self-esteem rises dramatically.

Self-esteem has been found to be positively correlated with achievement in the workplace and school. Covington's (1988) research showed that as self-esteem increases, achievement scores go up, and as self-esteem decreases, achievement scores go down. Furthermore, self-esteem can be modified through direct instruction, and such instruction can lead to gains in achievement. This indicates that achievement and self-esteem are reciprocal. Schools can teach self-esteem by placing responsibility on the students, providing more cooperative learning opportunities, instituting a peer helping service, and adopting a real-life curriculum.

Furthermore, Robert Reasoner, past president of the National and International Council for Self-Esteem, found in his work with high school students and instructors that when self-esteem was improved, attendance went up 70%, performance improved 50%, 30% more students graduated and passed competency tests, and 60% less used drugs on a weekly basis (Hawkins, 1994). Integrating writing and math in practical ways increases self-confidence and feelings of worth that affects students' lives in multiple ways.

Support From Professional Educational Organizations

Many of the standards set by leading organizations, such as the National Council of Teachers of Mathematics (NCTM), the International Reading Association (IRA), the National Council for Teachers of English (NCTE), and the Association for Supervision and Curriculum Development (ASCD), encourage integration of the curriculum, discovery learning, student learning groups, and purposeful, meaningful assignments, which are part of strategies described in this chapter.

The NCTM *Standards* (NCTM, 1989) applied the philosophy of making mathematics meaningful and personal by using writing and advocated the following precepts: (a) problem solving in its broadest sense is nearly synonymous with doing math, (b) basic skills in mathematics are more than simply computation skills, and (c) the success of student learning should be assessed by a wide range of diverse measures. Computational activities should not be taught in isolation from a context of application and called problem solving. Students should be encouraged to question, experiment, estimate, explore, and suggest explanations. Problem solving, as a creative activity, should not be built exclusively on routines and formulas. The mathematics instructor should assist the student to read and understand written problems, to hear and understand oral problems, and to communicate about problems in a variety of formats.

Developers of the IRA and NCTE language standards (IRA, 1994) are committed to the idea of integrating language arts as a tool for learning in all content areas. These Standards stress the importance of teachers knowing that language is an active meaning-producing activity that is a prerequisite for accomplishment in all subject areas and in many tasks performed outside of school.

Recommendations by the Association of Supervision and Curriculum Development/U.S. Department of Labor (ASCD-DOL) for strengthening occupational preparation of noncollege bound students suggest, among other things, that educators must "teach future employees how to make decisions, how to solve problems, how to learn, how to think a job through from start to finish, and how to work with people to get the job done" (Carnevale, Gainer, & Villet, 1990, p. 237).

The National Assessment of Educational Progress (NAEP, 1990) found that despite the emphasis placed on process writing in schools, recent national writing evaluations showed deficiencies with informative and persuasive writing assignments. Half the students in twelfth grade reported they had written no more than two papers for school during the previous six weeks. Students did not report that they consistently use the strategies associated with the process writing movement.

In summary, professional educational groups focusing on the teaching of mathematics, reading, supervision, and educational testing agree that the national curriculum should be more integrated, more student centered, and more applicable to real-life experiences.

Support From the Workplace

New strategies are needed to prepare students for the workplace. Elizabeth Dole (1989), then the Secretary of Labor, stated that the U.S. workforce is not ready for the new jobs and challenges of the 1990s. Jobs are demanding better reading, writing, math, science, and reasoning skills. There is a 25% high school dropout rate, totaling as many as one million a year. Seventy percent of high school seniors are incapable of writing a basic letter seeking employment information. Sixty percent of all high school seniors cannot compute their own lunch bills, and 20 to 40 million adults experience substantial literacy problems in the United States. The Motorola Corporation eliminates four out of every five applicants for entry level jobs by administering a test that covers basic skills from seventh grade English and fifth grade mathematics.

The ASCD and the U.S. Department of Labor (Carnevale, Gainer, & Villet, 1990) sponsored research regarding the technical workforce in the United States and the types of training they are receiving. Technical workers are people who use principles from the mathematical, physical, or natural sciences in their work to create products, services, or processes, and they form the backbone of American competitiveness in the world market (Carnevale, Gainer, & Schulz, 1990). ASCD-DOL found that the United States is not competitive with other countries in basic education and occupational training for noncollege bound youth, even though this group represents approximately 61% of the high school student population. Employers have been forced to remediate basic skills because labor shortages require them to hire young employees who formerly would have been considered unqualified. Additionally, advanced technologies require higher levels of basic skills (Carnevale, Gainer, & Villet, 1990).

> It is ironic that as the workplace becomes more technologically complex, the rising pool of available workers is lacking in many of the simplest and most basic skills, including reading, problem solving,

computation, and knowing "how to learn." In fact, 30 percent of those potential workers are likely to come from populations that are disadvantaged. (Carnevale, Gainer, & Schultz, 1990, p. 1)

Support From a Practical Viewpoint

Integrating the disciplines saves time. Goodlad (1984) stated after his extensive research on American schools that some schools seem incognizant of the fact that time is their most valuable learning resource. School-to-school differences in using time effectively make inequities in learning opportunities.

Using process writing to compose mathematical assignments calls for peer conferencing that empowers each student to fill the role of teacher for another student. It also allows instructors to be life-long learners, sharing in the exciting journey with the students and avoiding burnout. Instructors must not give up their legitimate authority and responsibility for guiding students. Rather, they must encourage students to develop their own ways of being responsible for learning (Wiske, 1994).

In summary, learning theories, professional organizations, and recent research support the need for teaching mathematics and language arts as integrated disciplines. Using process writing to teach mathematics is one practical, relevant, and motivational way to integrate the disciplines.

USING PROCESS WRITING TO INTEGRATE LANGUAGE ARTS AND MATHEMATICS

An uptown New Orleans GED class with students 22 to 67 years of age, made mathematics come alive by integrating math with reading, writing, and the real world. One specific instructional strategy used was process writing (see Atwell, 1987; Butler, 1984; Calkins, 1986, 1994; Graves, 1983). This approach involves cooperative learning, an equal role for student and instructor, and a specific timeframe and structure for writing, and may be referred to as the "Writers' Workshop." It consists of six recursive, overlapping stages:

- *Rehearsing* (brainstorming). This is the stage in which writers discover what they want to communicate to others (Temple, Nathan, Burris, & Temple, 1988). Graves (1983) called this being an observant liver of life. During this stage, encourage adults to explore their own thoughts and state opinions about the topic they have chosen. Jotting down quick titles, subjects, and relevant ideas at the time they are inspired helps students generate new pieces at a later time.

- *Drafting* (quickly getting initial thoughts written). The first draft is a preliminary version that may be revised many times before the piece is finished. Getting the meaning of the text across to the reader is the focus of this step. Students are not permitted to erase or to be concerned with the conventions of the language.

- *Responding* (to feedback based on reading and questioning from an audience). The response stage involves receiving feedback first from peers, followed by feedback from the instructor. The author reads the piece and literally "holds the pen" while the listener gives suggestions, asks questions, and seeks clarification. Responses concern only the meaning of the piece and begin with praise for the ideas expressed.

- *Revising* (re-seeing the piece and improving it). During revision for meaning, writers contemplate their compositions to see if they have said what they intended. They strive to make their ideas clear and logical to themselves and the reader. Writers decide if they want to incorporate any alterations suggested by peers or the instructors during the responding stage. When the meaning is clear, the piece is edited for the conventions of the language. Process writing is recursive, and revision often leads the writer to return to rehearsing or drafting of the same piece. Process writing values the journey through which authors travel to reach the final destination, the published product.

- *Sharing* (reading and receiving comments in large group). A sharing time at the end of each day's writing session is critical. Calkins (1986) calls this the Author's Chair. A good title or first sentence may be shared, and help may be solicited from the entire group. Sharing their works with others helps students value their own written thoughts and those of others. Sharing is also appropriate when the composition is completed.

- *Publishing* (printing the final product). After proceeding through the earlier stages in a recursive manner, students eventually have pieces of writing that are meaningful and conventionally correct. Publishing may be in the form of class or individual books, letters to the local newspaper, job mini-manuals, posters on the wall, or overhead transparencies to use for peer instruction. Author autographs and dedication pages are appropriate at this stage.

The next section describes the use of process writing to create original mathematical word problems. Students discover that active participation in the stages of the writing process encourages personal growth and self-knowledge, as well as facilitates the mastery of basic skills. Although the examples provided are from a GED class that met daily for three hours, this project can be easily adapted to adult basic education (ABE) classes, developmental college mathematics courses, workplace literacy workshops, and other mathematics classes.

Applying Process Writing to Original Math Problems

Setting expectations. The GED class involved in this project was typical in that students had diverse goals and needs. The class greatly enjoyed using process writing to create fiction, technical, and informational pieces and was now ready to write in a content area. The instructor introduced the idea of creating a book of original math problems, with each student writing three to five problems based on personal experiences. A library book was displayed with its various parts: the dedication page, table of contents, and title page. It was explained that the class book would have the same parts. These adults students were delighted with the idea. One woman explained when leaving that day that she was 66-years-old and had never before been an author.

Modeling an original math problem. In order to model expectations, the instructor wrote an original math problem from real life on the board and carried the students through the steps of reading it for meaning, revising, and editing. The rough draft, which included some words crossed out, read:

> I am baking. Last night I baked bran muffins for my family. The resipe called for: 2 cups cereal, 1 1/4 cup milk, 1 cup flour, 1/3 cup brown sugar, 2 tsp. baking powder, 1/2 tsp. baking soda, 1/4 c. melted margarine, 1/4 c. egg beaters, and fruit.

> Then, I noticed that this resipe only makes one dozen. I have 5 kids, and man can they eat! I better to double the recipe. Also, I didn't have the artificial egg beaters product. I decided to use one egg instead. After soaking the bran cereal in the milk for five minutes, mix all the ingredients by hand. Put in muffin tins and bake for 30 minutes at 350 degrees.

> How much of each ingredient will I need if I double the resipe?

The students were enthusiastic about sharing the everyday math problem with the instructor. When attempting to understand the meaning of the problem, one student asked what egg beaters were, so an additional explanation was added behind egg beaters reading "a liquid substitute available at the grocery." Other class suggestions included adding a subheading for ingredients and additional directions for clarification. Also, one student asked if the baking time would change if the recipe was doubled. Her reasoning was that if the muffins were larger, the baking time would need to be increased. The class concluded the problem by asking, "How much of each ingredient will I need if I double the recipe to yield four dozen muffins instead of two dozen?"

With the revised version of the demonstration problem still on the board, the students worked it, discussed it, and compared their answers. Differences between adding/subtracting fractions and dividing/multiplying them, as well as the terms numerator and common denominator, were discussed. Learning was taking place.

Agreeing on the answer on the board, the class looked back over the problem for editing errors including grammar, spelling, and punctuation. Knowing an interruption would interfere with ongoing thinking processes, the instructor initially circled "resipe" in the first sentence to signal that it should be looked up in the dictionary later. Now, after working on this one problem for 30 minutes, a student volunteer checked the correct spelling in the dictionary. "Resipe" became "recipe" and the challenge was given to the students to correct this spelling in other places in the problem.

Students' Journey with Original Math Problems

Rehearsal stage: Brainstorming daily math experiences. After working the instructor's problem just described, students were urged to go through the rest of the day or week, highlighting times they encountered mathematics in their everyday lives. Enthusiastically, they observed and considered their own individual lives complete with real problems that needed practical mathematical solutions. The resulting interaction was terrific. Even the quiet students hurried into the classroom each morning to discuss the new math problem they generated during the night. The topics included income tax refunds, wage raises at work, discounted prices of clothing, dental bills, and other personal finances. All ideas were accepted and praised, then followed with the command to "Write it so others can read it." These adult learners were on their way to becoming authors and pragmatic mathematicians.

Drafting stage: quickly getting thoughts written. Real-life math problems were written quickly without thinking about spelling or the conventions of the language. Erasures were not permitted because the meaning was stressed at this beginning point. Full of pride, one of the students brought the following rough draft and read:

My name is [X]. I ride the bus Mody, Wednesday, Friday to GED class at Trinity Episcopal Church on Jackson Avenue. I volunteer on Friday. I have been doing it for over a year. I like it very much and get along with the other workers very well. I ride the Bus every Mondy, Wednesdy Friday. How much is my bus I pay at 90 cents going and coming.

For week would be? For a month?

For three mons? For six moths? For a year?

Responding and revising stages: Reading and questioning from class-mates, and reviewing and improving the piece. The instructor immediately praised her for her story, particularly the volunteer work on Fridays. The instructor then showed interest in the meaning by asking, "What do you do on Fridays when you volunteer?" Next came clarification questions: "Have you been volunteering or coming to class for over a year?" "Do you pay 90 cents each way or for a round trip bus ride?" "How do you calculate the number of days you ride to school in a month?" The student author realized she needed to clarify these points in writing. After verbalizing her answer, she returned to her seat to make the need-ed corrections.

Following a peer conference to check the meaning and mathe-matical calculations, the student read the problem to the instructor again. The author then began editing the piece. The class calendar was used to check the spelling of the days of the week throughout the prob-lem. Next, the instructor circled the first spelling of month and asked that its spelling also be checked throughout the problem. After these cor-rections, the final draft was begun.

At this point in the writing process, the instructor typed the problem, double spaced, with the mistakes it still contained. Once the problem was typed and looked like a real text, it was easier for the writer to notice other errors. The student read it over one more time, making final corrections and additions.

This student's problem, as well as the problems of other class-mates, was fine-tuned for weeks during a portion of each class period. After the student wrote a rough draft, a checklist of conferencing steps was filled out. This included a place for the author, peer, and instructor to check questions regarding the content (meaning), mathematical soundness, and writing mechanics. Also, a peer worked the problem mathematically and checked the author's answer. The independence this generated gave each student, writer-reader, and peer editor, self-confidence as writers and instructors. Teacher conferences were held to listen to the rough draft only after the author and at least one peer read the piece, commented on it, and calculated the answer. Appropriate notations were placed on a checklist, as illustrated in Figure 10.1.

Sharing: Reading and receiving comments in large group. Throughout the writing experience, students participated daily in a sharing time with the entire class; they either read from their own writing, or listened and commented on the contributions of others. The following are some additional examples of original mathematical narrative problems provided by our students:

> After Jami figured out the income tax checks, she decided she wanted some new clothes. She went to Contempo Casuals in the New Orleans Center. Jami saw many things she liked. She saw 2 skirts at $18.00 apiece, and she saw three pairs of socks at $2.00 apiece. How much would Jami plan to spend if she decided to buy these items?

> To have a tooth put in costs $150.00. I have a receipt for $70.00. $80.00 is still owed on the ticket which I paid him. I left a ring for collateral. I did not pay any more money. How much do I still owe?

Some students got creative with dialect:

> If there's one thing that Jim Bob and Billy Boy are good at it's being lazy. They "wuz sittin' home one day watchin" "The Dukes of Hazard" on T.V. After it was over, they wanted to watch "The Fishin Show," but neither wanted to get up and change the channel. All of a sudden, Billy Boy gets an idea and wants to make a remote control switch outa a broom handle. If the couch is 7 feet away from the T.V., how long of a stick would he need if he moved the couch back twice as far?

Publishing: Printing the final product. After all students completed a similar procedure, the class was ready to put the final product together. The group decided on a dedication, wrote the table of contents, and put the answers to all of the problems in the back of the class book. A simple coil binding with a plain red cardstock cover and back was chosen at the local copy store. The book, entitled *Original Math Problems by Trinity Episcopal GED Students*, was ready to go to press.

Name _____   Date _____

EDITING CHECKLIST

CONTENT:	1st Draft			2nd Draft			3rd Draft		
	A	P	T	A	P	T	A	P	T
Is the meaning clear?									
What is confusing?									
Ways to be more descriptive									
General content additions									
FORM:									
capitals									
spelling									
whole sentences									
commas									
periods									
quotation marks									
paragraphing									
(mathematically correct)									

A= Author   P= Peer   T= Teacher

Comments (and date):   Author's Chair: _____
                                                  (Date)

Figure 10.1. Conferencing checklist

The final Author's Chair to celebrate the finished book was a tremendous success. For a few minutes, each student became a mathematical expert while reading, explaining, writing on the board, and fielding questions about his or her problem. It was evident that self-esteem had been raised through this instructional strategy. Eye contact improved and the self-confidence of these struggling readers and writers blossomed as they wrote and read their personal math problems. In the coming weeks, certain students became sought-after "authorities" on other problems concerning percentages, discounts, or decimals that were similar to the ones shared during Author's Chair.

The students autographed their pages in other people's books, writing comments to them. Each student had indeed become an author while learning a lot of math and language arts simultaneously. This type of follow-through and closure was very important to the success of the project.

Specific Aspects of the Writers' Workshop

Several specific instructional methods made the experience of the Writers' Workshop more meaningful. Next are brief descriptions of the timeframe of the Writers' Workshop (Atwell, 1987; Calkins, 1986; Graves, 1983), the mini-lessons, and the writers' roles (Flowers, 1981).

Timeframe. The Writers' Workshop refers to the framework of classroom time and philosophy devoted to teaching writing as a process. Graves (1983) recommended 45- to 60-minute sessions three times a week for the Writers' Workshop. The writing period begins with a 5-minute, instructor-directed mini-lesson. The content of the mini-lesson is driven by the mathematical or language arts needs of the students. Each mini-lesson may focus on one of several purposes: explain a procedure of the Writers' Workshop, model a mathematical computation, demonstrate examples of good literature, model peer conferencing, show positive revision strategies for editing, or teach a convention of the language (Calkins, 1994). An excerpt from a great literature book, the instructor's own personal writing, or a student's writing may be displayed on an overhead transparency. All mini-lessons end with a challenge for the students to apply the topic discussed that day. Some instructors ask the students to keep a spiral notebook with the date and the mini-lesson covered with an example written for future reference.

Most of the 35-40 minute writing period is devoted to actually writing and conferencing with peers and instructor. This encompasses the rehearsal, drafting, responding, and revising stages already discussed. The last 15 minutes of each writing session, Author's Chair, is crucial to improving the written work, as well as the self-confidence of the class. During this time, the student reads part or all of the piece aloud, calling on other students who have questions or recommendations. The instructor participates as part of the audience. If the piece is still being constructed, Author's Chair serves as a group peer conference, as well as a model of how to ask questions when the students are paired for one-on-one conferences.

Mini-lesson example. The instructor uses the following four-step plan to model mathematical computations or standard conventions of the language during daily mini-lessons at the beginning of the Writers' Workshop:

STEP 1: The instructor directly states the topic for the day.
STEP 2: The instructor and the class together correct an example that is shown on the overhead transparency.
STEP 3: The students write their own example. The instructor checks quickly for accuracy.
STEP 4: A challenge is given to apply this to writing or mathematical computation that day.

The instructor's part of a classroom dialogue of a mini-lesson on punctuation was as follows:

STEP 1: "I have been noticing something in your writing that we need to discuss today. When you have a long list of items, you put a comma between each of them to help the reader make sense of what you are saying. Look at this example on the overhead:"
'Jami saw many things she liked at the store. She saw skirts, purses, shoes, and scarves.'
"Notice that we put commas between the items on the list."

STEP 2: "Now, look at this example and tell me what this sentence needs:
'Often my son helps me around the house. He earns $5.00 for cleaning the gutters $10 for mowing the grass and nothing for washing the dishes.'
"Correct. We would put a comma after gutters and grass. Good!"

STEP 3: "Write a sentence on your paper that has a correctly punctuated series or list. I will come around and check it."

STEP 4:   "I challenge you today to look over your math problems to find sentences that need commas added to a series of words or phrases and to remember to add this punctuation for new writing as well."

Writers' roles. To make the writing stages meaningful to beginning writers, Flowers (1981) suggested a figurative way, in which students are given four role models: the madman, the architect, the carpenter, and the judge. The authors of this chapter feel that another role of social worker could be added (Hicks & Wadlington, 1994). Writers assume these roles as they work through the writing process.

The madman is in charge during the *rehearsing stage*. He or she has scores of ideas, writes spontaneously, and lets emotions run wild. The architect then comes in during the *drafting stage* and reads over what the madman wrote. The architect's task is to choose large sections of writing and *respond* by organizing them into a basic structure. At this point, the carpenter comes in to *revise* and build the piece further. The carpenter nails the ideas and sentences together, makes sure the writing is clear and sequential, and checks for transitions. The social worker *responds* by listening carefully when the piece is read, praises the author, and asks questions to clarify and extend the piece. The judge then takes over. He or she proofreads the whole piece for details such as grammar, tone, and spelling. The judge may send the piece back to the madman, architect, or carpenter *for more revisions*, or may declare the verdict that the work is finished.

OTHER APPROACHES TO INTEGRATING MATH AND WRITING

The following are brief descriptions of projects that many adult educators will find useful and in which process writing can be implemented to assist adult students in improving mathematical as well as literacy skills. These descriptions should serve only as a starting point to stimulate instructors to develop strategies that fit their own unique students and situations. (A growing number of publications also address writing and communication in the K-12 mathematics classroom and also present related ideas that are relevant to adult students, for their own learning as well as for helping them understand what their children may be doing in school).

Math Learning Logs

Students may keep math logs in which they record brief daily accounts of what they learned about mathematics with emphasis being placed on meta-learning. These entries can include short descriptions of class discussions that they found enlightening, their personal thought processes as they solved a problem, or reflections on their own mathematical learning styles. Students should be encouraged to consider their own mathematical progress and future learning goals. If students' entries are initially short and choppy, it may be helpful to give them specific open-ended questions until they become accustomed to expressing their ideas freely. Examples of questions are: "What did you learn today and what helped you learn it?" and "If you were going to teach today's lesson to someone else, what strategies would you use?"

Learning logs should be a risk-free activity. Emphasis should be placed on meaning rather than the mechanics of language, such as grammar, punctuation, and spelling. Logs should never be graded and only be shared voluntarily. However, if students choose to share logs, instructors can gain needed insight into students' thought processes, as well as the effectiveness of their own instructional techniques. Instructors may respond briefly in writing with meaningful positive comments and thoughtful questions.

Math Dialogue Journals

Math dialogue journals are different from learning logs in two important ways. First, dialogue journals explicitly encourage mathematical communication interactions between students and instructors. Students write entries keeping in mind that the instructor or another student will be reading and responding to their ideas. Usually, entries are informal with personal, conversational-like dialogue between the journal owner and the responder. Writers (journal owners and responders) must strive to state their ideas clearly and succinctly to make themselves understood. As writers clarify their ideas for others, the ideas also become much clearer to them. Often they recognize their own faulty thinking or unreasonable solutions to problems.

Second, although the focus of the learning log is on explanations of skills and concepts learned and on analysis of learning styles, the dialogue journal's content is much broader. Students may write about any aspect of mathematics that they wish to discuss with another individual. Popular topics may include applications of mathematics to everyday life, the beauty of mathematics, and mathematical successes and frustrations. Because students sometimes choose to discuss subjects that might be included in a learning log, it is best not to try to use dialogue journals and learning logs simultaneously. Instructors should choose the technique most suitable for their own students.

Some direct instruction regarding response techniques might be needed before students respond to each others' entries. No one should be considered an expert or superior to another. Responses should be honest but stated in a positive manner. Disagreements should be viewed as opportunities for growth, handled with respect for others' ideas, and always accompanied by written justifications. Emphasis should be placed on meaning and clarity; therefore, language mechanics should be stressed only as they relate to these. Students respond to mathematical ideas rather than correct entries for grammar and spelling.

This experience should also be risk-free and grade-free. Instructors should take notes regarding mathematical misconceptions and language errors for later instruction rather than use journals as didactic teaching tools.

Workplace Mini-manuals

Students can develop short, simple manuals that involve key aspects of their workplace activities. Depending on the situation, these mini-manuals may involve time schedules, payment plans, sequences of directions, listing of priorities, procedures for handling money, divisions of duties, estimation, exact accounts of costs and profits, and so on. Many students choose to write manuals that involve task analyses of their duties at work. For example, a table of contents for a fast-food restaurant cashier's mini-manual might include instructions for working the cash register or computer, directions for ringing up diverse orders, information about how to deal with customers, and procedures for clocking in and out.

It is important that students make their manuals mathematically accurate as well as edit for correct spellings, punctuation and grammar, and appropriate terminology so that the manuals may be understood by others. Illustrations, graphs, spreadsheets, and flow charts can also be used to clarify and explain. Color coding and indexing are helpful in organizing more complex manuals.

Mathematical Language Experience Stories

These stories are personal accounts of math happenings in students' lives and are especially useful with beginning readers. They can be introduced in small groups and later written individually. The group first selects a math experience that is common to the group, such as buying books or supplies, budgeting, home repairs, and so on. Each student then dictates one or two sentences describing the experience. The instructor records the story on a large chart in the exact language in which it is dictated, saying the words aloud as they are written. After the story is dictated, the instructor reads the story aloud, tracking the words with his or her hand. The group may decide to make revisions after this initial reading. After all revisions, the instructor reads and tracks the story again. Then individual group members read their sentences or all of the story aloud and later progress to silent rereadings. Because the experience is described in the group's own language, as they read, students are more likely to recognize sight words and decode other words using context clues.

Stories may then be used for follow-up activities that include additional silent rereadings, word recognition and comprehension activities, math skills practice, choral reading, taperecordings of oral readings, and so on. Students especially profit when they make personalized student word banks of new words (including mathematical terms) from their language experience stories. As students become competent and comfortable in dictating and reading group stories, they are able to begin to dictate their individual stories to recorders with the final goal being to independently write their own mathematical stories. Structured language experience stories in which students are instructed to use certain mathematical vocabulary terms may also be utilized.

Personal Math Histories

Students often enjoy writing their own personal math histories. In this activity, students record a chronological history of their own experiences with mathematics. These can begin with their earliest mathematical recollections (e.g., early birthday parties, trips to the store) and progress through their mathematical experiences in school (e.g., tests, instructors, successes, failures) up to ways they are involved in mathematics in the present (e.g., work, school, home). They should be encouraged to write about positive and negative experiences as they express feelings as well as facts. These accounts can be illustrated or put on timelines. For example, a student might choose to make a pictorial timeline of math experiences from birth to the present using photos, drawings, captions, and old school papers. Viewing and reading personal math histories helps instructors understand their students better and also assists in developing comradeship among students with similar experiences.

Math Creative Writing

Students can be encouraged to write creatively about mathematics. Their pieces can be very diverse. They can write short rhymes such as riddles or limericks or longer poems such as ballads. They can also create songs that may be set to traditional tunes or more modern rap beats. In addition, students may write short fiction stories that involve math skills and concepts.

It should be stressed that students need to hear poems, stories, and songs being read or sung aloud on a regular basis if they are expected to participate in this activity. (These do not necessarily have to have math themes.) Students will initially copy the techniques and phrasing of their favorite writers but later develop their own styles. As they start to connect mathematics instruction with creative outlets, they often begin for the first time to view math as fun and interesting. Creative activities become strong motivators for learning for many students who formerly had been insecure about their mathematical abilities and goals.

Creating Test Questions

Students enjoy writing math test questions. Guidelines must be set in the beginning and might include the following:

- Only ask about material that has been studied.
- Ask a thought-provoking question that makes use of words and numerals.
- Make sure your question is clear. Have someone outside your group read it to see if it is easily understood.
- Be able to show how you arrived at the correct answer. Check to see if more than one method or answer might be correct.

After the students create questions, instructors may choose to put all students' questions on the test or to create a test bank using only a few questions each time, making sure that every student regularly has a self-developed question on a test. The more instructors model good questioning techniques, the more likely they are to get good questions from students.

Newspaper Activities

Students may do activities that are centered around newspaper articles containing mathematical concepts. The sports and money sections espe-

cially lend themselves to math activities. However, shopping and classified ads, the weather report, and regular articles also are full of math
concepts.

Students may search for mathematical terms in headlines and
captions (e.g., "Airline Cuts Rates 50%!" or "City Loses 575 More Jobs to
Automation!"). Also, they may contact the writers of stories to discuss
mathematical issues or write letters to editors. In addition, they may
look for faulty statistical reasoning in persuasive stories, compute
gains/losses from the stock market report, plan activities using the
weather map, follow the lottery, or figure average salaries for different
jobs in the want ads. Often they are interested in analyzing sports statistics, planning meals using recipes and ads in the food section, or writing
advertising for items they would like to sell. Furthermore, they can
choose to write mathematical articles from different points of view or
purposely slant statistics to give support for opposing arguments. The
ideas for activities are limitless, and the payback is that students learn to
scrutinize media reports critically and carefully.

Mathematical Newsletter

A natural follow-up to newspaper activities is to create a class mathematical newsletter. Some students can write the actual articles for the
newsletter while others organize it into sections, do illustrations with
captions, design logos, write headlines, or do research. A distribution
list should also be established.

The newsletter should be very practical. Jobs available in the
area can be advertised with base pay and overtime salaries calculated. In
January, federal income tax information can be disseminated as well as
brief explanations of how to fill out necessary forms. Furniture and
clothing sales can be advertised with sales percentages deducted from
regular prices. Demonstrations of how to calculate local sales tax and
distinguish between items that are taxable and those that are not are
often popular. Information about social welfare reform bills before
Congress and their impact can be debated. Investment tips can be
shared. General information related to renters' rights are pertinent to
many adult students.

Cooking Experiences

Another practical application to the lives of adults is reading and writing recipes. After studying cookbooks from different regions, students
can write a favorite recipe or one that has been adapted for special needs

such as low salt, low fat, or diabetic diets. They can also explain how recipes can be doubled or halved, and costs/calories per serving can be estimated. A personalized, class recipe book of culinary delights can be the final product. A sampling of the prepared recipes is a welcome follow-up. (See the modeled cooking example in the "Writing Original Math Problems" section of this chapter.)

Math [History] Reports

One vehicle through which students may learn to view mathematics as a useful, fascinating body of knowledge is group report writing. Panels of students can research, write, and present reports on important people and events in mathematical history. Topics might include famous mathematicians such as Archimedes or Pythagoras; math created by different people (e.g., African nations), math in different eras such as the Middle Ages, the Renaissance, or today; careers in mathematics, or the development of a branch of mathematics such as trigonometry or geometry. Reports may be very elementary or more complex depending on the skills of the adult students. The instructor will need to provide research materials such as tradebooks, encyclopedias, and pamphlets if students do not have library skills or access to libraries. Nonfiction books in the young adult section of the library are often good sources, and many resources are mentioned in catalogs of mathematical associations such as the National Council of Teachers of Mathematics (NCTM). Using the computer to prepare reports can help make this activity exciting as well as encourage students to develop their computer skills in practical ways.

DISCUSSION AND IMPLICATIONS

The integration of writing with mathematics is advocated by numerous professional organizations, as well as by theoretical viewpoints related to cognitive and affective facets of the learning process, and is called for by the needs of modern workplaces. Process writing was positioned in this chapter as one key method to achieve such integration. The application of a process writing approach was illustrated earlier in the context of enabling students to develop original math problems, but it can be equally applied to the many other writing projects described previously.

Using process writing in any of these contexts initiates multiple processes that empower adults to not only develop new mathematical and literacy skills but to apply this learning to their lives in a meaningful, holistic manner. Construction of knowledge through active involvement can help learners become self-confident, competent mathematical problem solvers and communicators.

Our experience shows that during the implementation of process writing

- The role of the instructor and student merged as the community of learners-peers conferred, listened, and offered suggestions to others in the class.
- The social element of verbal interaction between students and instructor was crucial to the strategy's success. Students were interested in other people's writing (as well as in their own), encouraging and praising nonjudgmentally.
- The personal image of students was greatly enhanced. Choosing a topic and making decisions about a piece gave students a feeling of ownership. Author's Chair and peer conferencing built self-confidence and trust. Emphasizing meaning over mechanics assisted students in valuing individuals' own ideas.
- Mathematics and its real-life applications came alive for students, unlike story problems written in a workbook. Students began to value mathematics as a way to address life's issues and communicate about them.
- Students' literacy skills also improved. Concise meanings and correct language forms were strong goals because the book was going to be published. Students had a reason for writing clearly because they had an audience to read what they wrote.

Process writing requires that reading and writing be merged into a circular pattern of activities in which one leads to the other. When process writing is used for mathematics, students are able to transfer language arts skills to real-world tasks that require critical thinking and problem solving. This approach supports affective growth and empowers students to make choices and to feel ownership; consequently, it has the potential to greatly improve learning and adults' ability to handle life choices and workplace demands. Finally, this strategy is practical for instructors and learners. Integrating disciplines saves valuable time; it encourages students to be responsible for learning and helps students and instructors avoid burn-out.

In summary, writing original story problems and other projects benefits adult students in far more ways than indicated by improvement in math or in language skills alone. A simply constructed book of original problems, a personal cookbook, a mathematical journal, or a class newsletter become prized possessions and serve as a reminder of a time of active, meaningful, and comprehensive learning.

REFERENCES

Applebee, A. (1981). *Writing in the secondary school: English and the content areas.* Urbana, IL: National Council of Teachers of Mathematics.

Atwell, N. (1987). *In the middle: Writing, reading, and learning with adolescents.* Portsmouth, NH: Heinemann.

Butler, A. (1984). *Towards a reading-writing classroom.* Portsmouth, NH: Heinemann.

Calkins, L. (1986). *The art of writing.* Portsmouth, NH: Heinemann.

Calkins, L. (1994). *The art of teaching writing.* Portsmouth, NH: Heinemann.

Cambourne, B. (1988). *The whole story: Natural learning and the acquisition of literacy in the classroom.* Auckland, New Zealand: Ashton Scholastic.

Carnevale, A., Gainer, L., & Schulz, E. (1990). *Training the technical workforce.* San Francisco: Jossey-Bass.

Carnevale, A., Gainer, L., & Villet, J. (1990). *Training in America.* San Francisco, CA: Jossey-Bass.

Covington, M. (1988). Achievement dynamics: The interaction of motives, cognitions, & emotions over time. *Anxiety Research, 1*(3), 165-183.

Cunningham, J., & Ballew, H. (1983). Solving word problem solving. *The Reading Instructor, 36*(8), 836-839.

Dole, E. (1989). *State of the workplace address.* Paper presented to the state teachers and principals of the year, Washington, DC. (Eric Document Reproduction Service No. ED319967)

Flower, L. (1987). *The role of task representation in reading to write* (Tech. Rep. No. 6). Berkeley: University of California, Berkeley, Center for the Study of Writing.

Flowers, B. (1981). Madman, architect, carpenter, judge: Roles and the writing process. *Language Arts, 58*, 834-836.

Goldberg, N. (1986). *Writing down the bones.* Boston: Shambhala.

Goodlad, J. (1984). *A place called school.* San Francisco: McGraw-Hill.

Graves, D. (1983). *Writing: Instructors and children at work.* Portsmouth, NH: Heinemann.

Harste, J., Short, K., & Burke, C. (1988). *Creating classrooms for authors.* Portsmouth, NH: Heinemann.

Hawkins, J. (1994). Is there a correlation between self-esteem, achievement, and behavior? *The ASCA Counselor, 31*(4), 16-17.

Hicks, K., & Wadlington, E. (1994). Making life balance: Writing original math problems with adults. *Adult Learning, 5*(5), 11-13, 27.

Hiebert, J., & Carpenter, T. (1992). Learning and teaching with understanding. In D. Grouws (Ed.), *Handbook of research on mathematics teaching and learning* (pp. 65-97). New York: Macmillan.

Holdaway, D. (1986). *The pursuit of literacy: Early reading and writing.* Dubuque, IA: Kendall Hunt.

International Reading Association. (1994). *Draft standards.* Newark, DE: Author.

Jacobs, H. (1989). *Interdisciplinary curriculum: Design and implementation.* Alexandria, VA: Association for Supervision and Curriculum Development.

Kohn, A. (1993). Choices for children: Why and how to let students decide. *Phi Delta Kappan, 75*(1), 9-16.

National Assessment of Educational Progress (NAEP). (1990). *Learning to write in our nation's schools: NAEP's 1990 portfolio study.* Princeton, NJ: Educational Testing Service.

National Council of Teachers of Mathematics (NCTM). (1989). *Curriculum and evaluation standards.* Reston, VA: Author.

Newkirk, T., & Atwell, N. (Eds.). (1992). *Understanding writing: Ways of observing, learning, and teaching.* Chelmsford, MA: Northeast Regional Exchange.

Routman, R. (1988). *Transitions.* Portsmouth, NH: Heinemann.

Silverman, F., Winograd, K., & Strohauer, D. (1992). Student-generated story problems. *Arithmetic Teacher, 39*(8), 6-12.

Temple, C., Nathan, R., Burris, N., & Temple F. (1988). *The beginnings of writing.* Newton, MA: Allyn & Bacon.

Tierney, R., Soter, A., O'Flahavan, J., & McGinley, W. (1989). The effects of reading and writing upon thinking critically. *Reading Research Quarterly, 24,* 134-169.

Weaver, C. (1988). *Reading process and practice.* Portsmouth, NH: Heinemann.

Winograd, K. (1993). Selected writing behaviors of fifth graders as they composed original mathematics story problems. *Research in the Teaching of English, 27*(4), 64-67.

Wiske, M. (1994). How teaching for understanding changes the rules in the classroom. *Educational Leadership, 51*(5), 19-21.

III

REFLECTING ON PRACTICE AND LEARNING

11

Learning to Learn: Mathematics as Problem Solving

Leslie Arriola

This chapter describes my teacher-as-researcher project, which investigated, in one ABE math class, the development of "learning to learn" skills in a classroom environment based on implementation of the NCTM *Standards*.

INTRODUCTION

In the Fall of 1992, as a member of the Massachusetts Adult Basic Education Math Team, I became involved in the Team's challenging project to adapt and implement the National Council of Teachers of Mathematics (NCTM) (1989) K-12 *Curriculum and Evaluation Standards* in adult basic education (ABE) settings (see chap. 3 for a review of the *Standards*). Our project, which was supported by a federal grant, challenged our group of over 20 educators to articulate clearly the ways in which the needs of adult mathematics learners are similar or different from those of their K-12 counterparts. For this, we drew on our teaching experience and observations. Implementing the *Standards*, however, challenged us to explore a teacher-as-researcher approach to teaching—to methodically investigate, first hand in our classrooms, the ways in which *Standards*-based teaching might impact on our adult students.

As we attempted to characterize our adults' strengths and weaknesses as mathematical learners, it became increasingly apparent that before we could talk about implementing the *Standards* in our ABE classes, we needed to define more clearly the barriers to learning that adults face in the mathematics classroom. In this context, the question that interested me was not "What don't they know?" but, more fundamentally, "What is it that makes learning math so difficult for adults?"

The work of Dewey and Piaget has given us a compelling theoretical basis for our current view of knowledge and our beliefs about learning. Implicit in the *Standards* is the Piagetian view (cited in Ginsburg & Opper, 1979, and see chaps. 2 and 3, this volume) that knowledge is constructed through action and experience, and the belief that learning is a process of developing powerful thinking tools for understanding and solving problems in the real world. Implementing this view means creating a learning environment in which students are active agents in their learning process and in which curiosity, risk taking, inventiveness, and reflection play major roles.

We teachers think this learning environment provides an exciting and effective way to learn (and to teach)—more productive than traditional, passive learner modes, more in keeping with what we have learned about cognitive processes and how people acquire knowledge, and certainly more fun. From our perspective, it seems reasonable to assume that by simply removing the barriers and judgments that perpetuate passive learning, our adult students will move naturally and with great relief into an active learner role. But, in fact, our ABE students haven't the slightest idea what it means to learn math in this way. This is all new territory for them, alien and often threatening.

Our adult students do not understand what we mean when we say we want them to be mathematically curious. It is beyond their realm of experience to learn to think mathematically without an expert to tell them what to know and how to know it. They do not know how to reach for the knowledge, to look inside their own minds for new ways to look at math. They have never learned to value the struggle as part of the process. In math, at least, they have not yet learned how to learn. The situation is not due to some deficiency in adult students but is instead a result of their past experiences (which are often negative) with traditional math instruction. This gap in experience and understanding, I have come to believe, is the underlying problem teachers face in a *Standards*-based approach to adult mathematics learning.

This chapter describes my teacher-as-researcher project, which investigated, in one ABE math class over a period of four months, the development of "learning to learn" skills in a *Standards*-based classroom that focused on learning fractions. The reader should keep in mind that, although there are many interesting questions to pursue about the methods and activities I used for fraction learning, the primary focus of the project and this chapter is on the general issue of developing adult students' math learning skills.

LEARNING TO LEARN

The vision of mathematics learning in the NCTM *Standards* implies a significant departure from the traditional practices of mathematics teaching. Reaching for the goals articulated in the *Standards*—that students learn "to value the mathematical enterprise, develop mathematical habits of mind, and to understand and appreciate the role of mathematics in human affairs" (NCTM, 1989, p. 5)—means changing students' beliefs about what it means to learn and do mathematics. It also means changing the roles that teachers and students play in the learning process. Getting students to believe that doing mathematics involves reasoning, analyzing, problem solving, and communication, not just computation and following algorithms, is one part of the challenge of implementing this vision. The other part, the part that invariably holds up the process, is the challenge of getting students to take a far more active role in the way they go about the learning process itself—about learning to learn.

Learning to learn means taking the reins in the pursuit of knowledge, rather than waiting to be led. It means that, faced with a problem or a new concept, students will explore, question, hypothesize, and struggle for understanding on their own, and that they will gain skills and confidence as mathematical investigators.

In my classes, I struggle constantly with students' resistance to becoming active, independent learners. I know this resistance is, in part, because I am asking them to play a role they do not understand or for which they have no experience. But it is more complicated than that. The following situation is an example.

Using fractions, Judy is trying to figure out how she spends her time on a normal weekday. On her own, she has made a list of activities with estimates of the hours per day she spends doing each. Looking over her shoulder, I see she has begun to translate hours into fractions:

sleeping - 7   7/24
working - 8   8/24
cook/eat - 3   3/24
driving - 1
chores - 1-1/2

Wondering what she will do with the 1-1/2 hours for chores, I sit down beside her and ask her to explain what she has done so far. She looks at me in a sudden panic: "This is all wrong, isn't it?" "How did you get the 7/24?" I ask, thinking that as she explains it to me, she will regain some confidence in her work. She searches my face for clues and says, "7 out of 24 hours . . . but now that I look at it, maybe it should be 7/17 . . . 24 minus 7 is 17." I keep quiet. She erases all the fractions and turns to me, defeated and expecting me to tell her what to do.

We have all been there. The minute we sit down with students, they stop thinking. It is an automatic reaction, a conditioned response: if the teacher asks you why you did something, it must be wrong. The answer is wrong, the reasoning is wrong, abandon ship—no matter how much sense it made before. Outside of school, Judy is a decisive, confident problem solver, but when it comes to doing math in a school setting, she has no confidence in her thinking. In part, Judy, and many ABE students like her with whom I have worked, expect to repeat their past failures in math. I believe this in large part stems from the students' deep-seated belief that math students can only know what their teachers tell them to know, and math students can never know if they are right until a teacher says so.

Then there is the interaction between students' beliefs about mathematics and assumptions about the learning process, as the next situation illustrates.

My graduate equivalency diploma (GED) math class has been making wall-size fraction bars out of 15-foot lengths of butcher paper sheets for a game we are inventing. They are working in pairs. The 1/3 and 1/5 people are moving right along, but the 1/4 and 1/6 ones have hit a snag: 4 and 6 do not divide evenly into the 15'. They do not know what the remainder means in terms of feet and inches. They ask me what to do. A great sidetrack, I am thinking.

I ask them what they think a remainder means. They look at me blankly. I pull out a set of unifix cubes. Work with your partner, I say, and figure out how to show with the cubes 15 divided by 4 and 15 divided by 6. The room is suddenly silent except for the sound of each student wanly moving his or her cubes around on the table. After 30 seconds, even that sound is gone. I circulate, ask a few questions. It is obvious they do not know where to begin and it has not occurred to them to experiment. They are sitting next to their partners, but they are not talking to each other.

Their dependency on me, the "authority", is all too typical in ABE math classes. Adults who are normally exploratory and inventive in how they solve everyday problems seem to leave these reasoning tools at the door of the mathematics classroom. Their view of math does not include invention and exploration. Math is cut and dried, something that has rules and methods that have to be memorized. And their view of math certainly does not include socializing. Talking to other students about how to do a problem means you are not smart enough to solve it on your own. If you're stumped, ask the teacher. If you want to learn something new, ask the teacher. If you want to know if you're right, ask the teacher.

When I try to "teach" less—more open-ended questions, fewer answers, less lecturing—because I want students to do more, they become confused and often panicked. They go blank. Over the years and with various groups of adult learners, I have tested with some success a variety of tactics to encourage students to take hold of the reins of their learning process and to change their beliefs that they are powerless as learners. But all too often, in the face of their confusion and passive blankness, I find myself back at the front of the class, trying to find a more comfortable launching pad for learning independence. And wondering what I could do differently.

Learning to Learn Takes Time

Students' belief that math is a mystery on which only some authority can shed light is deeply entrenched. Changing this belief is not easy. Getting ABE adults to change the way they go about learning math is even harder. Both take time. But how much time?

Our project gave me the opportunity to take a closer look at the process by which students develop into independent learners and to question my expectations as to how long this process might take. My own project involved taking time and tossing it out the window. Or, rather, gathering up all the time in the world and giving it to my math students. I wanted to give them a no-stress, no-tests, no-time-constraints learning environment, one in which they (and I) could take whatever time they needed to discover and develop their abilities to be active, independent, self-monitoring learners and knowers.

Knowing that weaning students from their dependency on me is a complex process, I explored a variety of methods to move us out of the traditional roles they expected and into new, more independent roles. I tried to go at their pace, not mine, in the hope that their resistance to change would fade pleasantly away, rather than having to be overcome. And I tried to make it all fun.

MY CLASSROOM PROJECT

My class for this project was made up of a small group of students recruited from traditionally taught Pre-GED and GED classes at a local adult learning center. These were students who wanted additional help with their math skills. A few of the students who signed up for the class had been in a previous "Math Fun" class I had taught at the center, but to most of them I was a stranger and my teaching style equally strange. The six students who formed the core group were diverse in both cultural background and mathematical achievement. All had had some experience learning math, but none of them felt they had a real understanding of even the most basic math concepts.

My offer to teach at the center happily coincided with the eagerness of David, the director and head teacher at the center, to find new and better ways to teach math to his students. Fortuitously, David became my partner in this project, not only in working with the class, but as a co-investigator. Although David had some awareness of the ideas in the *Standards* and had learned a little about manipulatives and active learning methods at staff development math workshops, he taught math in the traditional way—by helping students work through a series of GED-level math workbooks. He felt some students were learning math this way, but that "it was a drudge." It had also become clear to him that many of his students had deep-seated conceptual blocks for which the workbook approach was frustratingly inadequate. Through observation and collaboration, he hoped to gain new methods and a bigger framework to guide him in changing the way he teaches math.

David's interest in exploring new conceptions through my teaching meant that not only would I have the opportunity to implement the *Standards* in a situation free of time or curricular constraints, I would have a chance to introduce *Standards*-based perspectives and methods to my students' regular teacher. I hoped the spirit and philosophy of the Standards would provide David with the framework for change he so eagerly sought.

The class met twice a week. I taught the lead lesson on Wednesdays with support from David, and on Mondays David followed up on the lesson alone. Some students came to both classes, some to one or the other.

The Process: "Mucking About" in Fractions

David and I decided to focus on one mathematical concept for as long as it would take for students to get a solid grasp of it. We decided to center the lessons around a non-computational exploration of fraction concepts. Revisiting each concept often, but with different representations and in a variety of contexts, my goal was to immerse students in fraction *explorations,* instead of leading them through the traditional sequence of fraction topics and computations.

In four months of "mucking about", we:

- Divided things (wholes) up and put them back together again
- Played with different referent wholes
- Investigated improper fractions, proportions, and fractions of fractions
- Compared fractions of discrete and continuous quantities
- Worked in both abstract and practical contexts

I asked students to work on fraction activities and problems in pairs and small groups, at their desks, at the board, around the room, and even out in the parking lot. We made lots of manipulatives, invented games, made charts, drew diagrams, made up word problems, compared solutions, and talked. We talked a lot—to clarify, argue, question and pinpoint confusions. And we laughed a lot.

Every Wednesday, I would invent an activity to start off the class, which I would outline to David a few minutes before class. There was always a feeling of adventure as I began each class, partly because most of my starting points were just as new to me as they were to the class and partly because I knew I could never predict the students' willingness to participate.

During class, David took notes on what transpired: what I said and did; student responses, confusions and "aha" experiences; where we digressed or got bogged down. Also during class, he helped monitor small-group activities and lent his good humor to the "fun" atmosphere I was trying to create.

After each class, David and I would talk about how it went: What were the high points? Were we finally seeing some signs of independent thinking? Leah was on a roll today. Who got lost? How can we keep Oeuy on track? Why does Jackie refuse to work in groups? What did we (David and I) learn? What would be interesting for David to follow up with on Monday?

On Tuesday, David would briefly fill me in on what he and the students had done on Monday and how it went. Taking my lead from what he reported, I would then devise a plan for the class the next day. Sometimes there was continuity between our activities on the different days, sometimes we took off in different directions.

For me, David's notes, observations, and collaboration were a gold mine. How wonderful to have another set of eyes and ears to help me see what went on during the class. How helpful to have someone who knew each student's background and problems help me interpret reactions and responses. What a treat to have someone with whom to bounce ideas back and forth.

WHAT DID I DISCOVER?

First, I discovered that without the pressure of time and tests, students were quite happy to, as David put it, "swim around" and around and around, week after week, in noncomputational fraction explorations. They took readily to hands-on learning: folding and cutting up squares and rectangles, exploring pattern blocks, making color-coded decks of fraction parts to "pack" and "unpack" into wholes and groups of wholes, and finding the fractions in the room, in their homes, and in the world.

Second, timelessness gave us space to unhurriedly sidetrack into gaps and confusions whenever they came up: What's a remainder? How can you divide a big number into a small one? How do you make out a check? How does addition relate to multiplication? These were wonderful sidetracks, which more than once led us deep into unexpected investigations.

One day, for example, I began by asking students to look around the room and find fractions. Cass, grabbing a felt tip marker, said: "It's 1/8 red on the outside." "How long is the red part?" I asked. "The marker is 5" long," Cass said as he measured it. "Then," he declared, "the red part is 1/8 inch long." And off we went on a wonderful adventure to sort out the difference between 1/8 of an object and the actual measurement of 1/8 of 5". There was no way I could anticipate all of his gaps and confusions, but here we had the luxury of being able to take all of the time we needed to deal with them when they came up.

It was not long before it was clear students were really getting a feel for fraction concepts. They were also becoming more comfortable with my untraditional manner and approach. They answered my questions more readily. They were less shy about going to the board. They talked more, although still mostly to me rather than their peers. They let me know when something suddenly made sense to them. And, of great interest to me, confidence ratings were higher. I would often push students to commit to their answers by asking: "On a scale of 1 to 10, how sure are you of your answer?" They were beginning to know when they knew.

In fact, after just the third class, I wrote in my journal:

> This was one of those perfect classes that was full of "teaching moments." The class went so well the students actually clapped at the end of it and many told me how clearly they had understood the concepts we were exploring—and how much fun it had been. I confess to having had a wonderful time as comic orchestrator of the proceedings. . . . I felt so "right on" the whole time . . . guiding the activities, drawing them out to conjecture, take risks, explain, make up questions . . . making it funny and fun. We were all motivated, engaged, energized.

"What a high! It can't get any better than this," I remember thinking as the class ended. It was only later, when I took a longer look at the day, that I reminded myself in my journal:

> But . . . this was a class with the teacher at the front and in charge, and the students still dependent on her to lead them.

I am at home in front of the class. Like so many teachers, I am a good explainer. I ask good questions. I can draw students out and keep them moving. I am challenging, often funny. I was working hard for them and they were working hard for me. And that was the problem. They were still working for *me*.

As the weeks went by, it still seemed that without me to push and prod and orchestrate, they were lost. If, for example, I asked them to draw a picture of 12/5 of a candy bar, they would give up before they really tried and wait patiently for me to rescue them, no matter how long I waited them out.

When I would ask them to work in pairs, it was all I could do to get them to move closer to their partner, much less work on the problem together. They were lively, attentive and compliant—and still apparently totally dependent on me. When they came up blank, was I jumping in too soon? Did I need better questions? How much should I let them struggle? How much could I expect them to discover on their own? How much did they need me?

Learning to Change

I kept trying. I stayed alert for the smallest sign of independence and reinforced it. I moved them to one big table we could all fit around, which improved group discussions. I got better at devising activities in which they could not avoid working together. For example, asking them to measure the room with a long string meant they *had* to ask someone else to hold the other end and giving each pair only one piece of paper to record their measurements forced them to work together to come up with a relative unit (e.g., one string length equals one arm span) of measurement. And I kept learning to reframe my questions. By the third month they were clearly more knowledgeable about fractions and talking to each other a little more.

But were they more independent and self-reliant thinkers yet? At the time, I did not think so. Then, as I read through David's and my notes, I found there were more signs of progress than I had realized: Dave (one of the students) was making up a new variation of our fraction game; Jackie was monitoring and describing the way her mind works when she does math; Silvina was realizing she needs to listen better if she wants to learn better; Silvina and Oeuy were "arguing" over the way to approach a problem; and a wonderfully inventive fraction treasure hunt they concocted for me (with David's guidance).

I looked for more signs and they were there. They were beginning to work longer at difficult problems. They were taking more risks in their efforts and were more confident when an effort made sense. On their own, they were noticing connections, seeing patterns, finding relationships. A signpost I began to notice with great pleasure was that I was structuring the plans for each class less and less. That is, I could come in with just a starting point for investigation—for example, how to allocate the spaces in the center's parking lot (what fraction for students, for teachers, for deliveries, etc.?)—and be reasonably confident that they would become engaged with each other in the discussion without a lot of lead-up activities initiated by me.

David's report that, in his Monday class, students working in pairs to find the cost of a recipe actually worked together was music to my ears. In my class, too, I realized there was no longer that passive, waiting silence when I left them on their own with a problem. Here and there I could hear the sounds of minds thinking and communicating about math. Now, when I said to them, "The answer is 7/16. Make up a question," they had a better idea of what to do. With less and less help, I was getting better and better questions. They were arguing with me now and then, explaining to each other, pointing out mistakes.

Yes, I am doing less and the students are doing more. But we have just begun the shift. They are learning to learn, but progress is slow and these moments of independence and self-reliance are random and inconsistent. Yes, I am teaching a little less (i.e., I don't hear the sound of my voice quite as much), but there is still much to do to keep the process going.

As we continue to meet, the questions I ask myself as I plan for the class are more developmental: Where or what is the next challenge they can handle? How much longer should I leave them to struggle with a problem than I used to? How much more ambiguity can they deal with now? How much less do I have to spell out for them? How much are they ready to explore on their own?

DISCUSSION AND IMPLICATIONS

The structured inquiry I was able to conduct has helped me identify certain aspects of the process of the development of ABE students' learning to learn skills and dispositions, but it has also raised many perplexing questions that we can at best only partially answer. Next I first discuss points and raise further questions related to learning to learn issues and then note some observations that extend beyond the scope of my initial questions.

The time frame and context for this research project were unique in ABE settings. Rarely do teachers have the luxury of unlimited time to let students go at their own pace and to encourage sidetracks into areas of interest and speculation. It may thus seem to some readers that, whatever ideas about learning to learn may have emerged from my project, they may not transfer to regular ABE situations. But, is timelessness in a *Standards*-based classroom a luxury or a necessity? What are we saying to our students when we tell them, "Explore, investigate, reflect," and then add "and be quick about it?"

I am now more aware of the fact that in ABE mathematics classrooms, theory and practice do not always coincide. In theory, an adult's real-world problem-solving skills should transfer nicely to an active-learning, problem-solving mathematics environment in the classroom. When, in practice, adults turn out to not exercise real-world skills they may possess in our *Standards*-based classes, it behooves us to take a closer look at beliefs and expectations—theirs and ours—about what it means to be a student and what constitutes successful learning. When students cling to a passive learner role and discount their own reasoning abilities, what do we need to do to dispel deeply ingrained expectations that the teacher's job is to give knowledge and the student's job is to receive it? How do we convince adult learners that their ability to reason is a valid and necessary part of their learning process?

This teacher research highlighted the need to question the generality of our assumptions about adult mathematics learners. After three months of fractions activities, these students were only beginning to understand some of the more abstract fractions concepts. We cannot assume that the ability to think abstractly is invariably age-related. Nor can we assume that because an individual can think abstractly about human psychology or religion, he or she will be able to immediately grasp the abstractions in mathematics or physics. Can we expect someone, just because he or she is an adult, to learn, in two or three months of two-times-a-week classes, the mathematics thinking it takes younger students years to develop?

More related to the nature of learning to learn skills and dispositions, I learned to remind myself time and again that because learning to learn is a slow process, teaching less will also be a slow process. Just because ABE students are adults we cannot expect them to become active learners the minute we give them active-learning experiences. People do not become analytical overnight. Beliefs do not change overnight. The ability to communicate one's reasoning does not come easily.

I have learned to pay closer attention to students' starting points. If they seem passive and dependent, I have found it is often because the idea of, for example, exploring mathematical relationships is meaningless to them. It has no basis in their experience. Nor can I expect them to defend their reasoning when they do not have the mathematical language to describe what they did. I am more aware now that not all of the mathematical tools I want them to acquire can be discovered. Some things I have found I must tell them, model for them, or guide them toward.

I am more certain now that giving students this beginning support is not a sign of failure, on their part or mine. To the contrary, it simply signals the beginning state of a learning collaborative in transition. It is also a realistic indication of how long it may take to shift the expectations of students who complain, "Don't ask me to think, just tell me the answer," from wanting answers from the teacher to demanding that they have a chance to make sense of a situation themselves. In time and with practice—lots of time and lots of practice—they will go looking for their own mathematical tools, for new ways to approach problems, and for new ways to reason mathematically.

Finally, this project also highlighted certain issues about how teachers' (rather than students') professional knowledge develops.

The opportunity to team teach a class which was afforded to me is rare in Adult Basic Education. Yet, the benefits of this teaching partnership to the teachers (as well as the students) proved to be significant and indicate a need to raise questions at the organizational level about effective teaching practices.

More broadly, adopting a teacher-researcher role proved beneficial for both teachers and students in this project. Teachers and educational researchers have traditionally been separate entities. The researchers pose the questions, conduct the investigations, analyze the results, and present their findings to the educational community to interpret as best they can, leaving teachers to second-guess the relevance of the research to their students in their teaching centers.

In contrast, teacher research, such as I have described in this chapter, is always relevant to the teacher who conducts it. (See also chap. 5, regarding steps teachers can take to implement and explore the impact of new practices.) Teacher research turned out to be an approach that, when applied on a regular basis, keeps attention focused on what we can learn about our students from the students themselves, and what we can learn about ourselves as teachers from reflecting methodically on our teaching practices and classroom experiences, thus providing a concrete basis for evaluating and improving teaching and learning.

REFERENCES

Ginsburg, H., & Opper, S. (1979). *Piaget's theory of intellectual development.* Englewood Cliffs, NJ: Prentice-Hall.

National Council of Teachers of Mathematics (NCTM). (1989). *Curriculum and evaluation standards for school mathematics.* Reston, VA: Author.

12

Journey Into Journal Jottings: Mathematics as Communication

Donna Curry
Massachusetts ABE Math Team

This chapter describes attempts to implement one of the NCTM *Standards*, that of "mathematics as communication," in an adult workplace education classroom. The chapter focuses on the impact of introducing journal writing on students' learning processes and reflects on some of the issues that come up when an adult educator attempts to reflect on and change his or her practice.

INTRODUCTION

In 1992-93, as part of a federal grant, the Massachusetts Adult Basic Education (ABE) Math Team revisited the National Council of Teachers of Mathematics *Curriculum and Evaluation Standards* (NCTM, 1989) and adapted them to adult learners. Based on the NCTM suggestions four overarching themes, or process standards, became the heart of the

Massachusetts ABE Math Standards: Mathematics as Problem Solving, Mathematics as Communication, Mathematics as Reasoning, and Mathematical Connections. One part of the grant stipulated that Math Team members would have to implement and document some aspect of the revised standards. I chose Math Standard 2: Mathematics as Communication to consciously implement and document in my classroom, and focused specifically on journal writing as a means of communication in a workplace math classroom. (See chap. 3 for further information about the NCTM *Standards*; chap. 13 elaborates on broader issues involved in developing communicative and literacy skills in an adult math class.)

For the past three years, as part of my work in workplace training classes, I worked closely with a teacher who had been extensively trained in the "whole language" approach. She is very knowledgeable and innovative in curriculum areas involving language. However, she is a math "phobic," having received very little math training even though she is certified in elementary education. In one of our many discussions about this lack of math background, I questioned why she had such vast exposure to whole language yet so little in math. She explained that she was mentored by a whole language "guru" who believed that the whole language process could be used in any area of the curriculum—any area except math.

This bothered me tremendously. I believe that all areas of the curriculum can and should be integrated and that language flows throughout everything we learn and do. This belief, and the assumption by so-called educational experts that math should be treated differently from other curriculum areas, spurred me to design my research project around the use of language in my math class. I believe that "communicating" in math is an important step in learning math, so I decided to look more closely at the Mathematics as Communication standard:

Standard 2: Mathematics as Communication states that curriculum should be designed to allow the learners to:

- Develop appropriate reading, writing, listening, and speaking skills necessary for communicating mathematically in a variety of settings
- Discuss with others, reflect, and clarify their own thinking about mathematical outcomes, and make convincing arguments and decisions based on these experiences
- Define everyday, work-related, or test-related mathematical situations using concrete, pictorial, graphical, or algebraic methods
- Appreciate the value of mathematical language and notation in relation to mathematical ideas.

I chose to explore the impact of using journal writing as I felt it would address the writing skills for communicating mathematically and provide opportunities for students to reflect on and clarify their own thinking about math. I also felt that journal writing would be fairly easy to implement into a rather structured class. (Because students attended my class during factory work hours, it was important to have a "structured" class so that supervisors would not perceive the course as a "waste of time.")

The key question I chose to consider was, Does writing about what is being learned in a math class lead students to a better understanding of and broader insight into math concepts? In this context, I also wondered if students would become more confident in math by writing about what they were learning and more comfortable "talking" the language of math as they wrote.

MY CLASS

The class, Prep Math, is taught two times a week for 1-1/2 hours, over nine weeks. It is the final class in a series of four courses designed to bring employees from a very limited understanding of math up to a level at which they would have adequate math skills to successfully pass the math portion of the Certified Quality Technician (CQT) test.

The CQT exam is a nationally recognized standardized test used by the American Society of Quality Control (ASQC) to recognize individuals as having basic knowledge about quality concepts. The company in which I was teaching encouraged employees to become certified as quality technicians; the national recognition would lend credibility to the individuals as well as to the company.

As is typical of workplace education classes in the United States, this class was driven by several agendas: the company was looking for employees who could become nationally certified; employees had their own agendas that often revolved around them becoming better able to maintain their position within the company. My own agenda as an instructor was to help students become more comfortable with tackling any kind of math problem and more willing to learn and attempt new strategies. I also wanted them to be able to apply what was being taught not only during the CQT test but on the job and in their daily lives. Last, I wanted them to enjoy the experience of learning math.

Employees who participated in the Prep Math class had either completed the other three math classes or had successfully passed a teacher-made assessment. Employees entering Prep Math already had some exposure to fractions, decimals, percents, problem-solving strategies, algebraic notation, and Pareto charts and histograms. In my Prep Math class, they were to cover the following topics:

- Notation (algebraic and scientific)
- Metric system (including metric and SI conversions)
- Geometry (perimeter, area, volume)
- Probability (including a brief introduction to permutations and combinations because sampling in the workplace is based on these concepts)
- Histograms, measures of central tendencies (mean, median, mode), and measures of dispersion (range and an introduction to standard deviation).

MY RESEARCH APPROACH

The research project, to be implemented in the context of an actual workplace classroom environment, consisted of three elements: journal writing, one-on-one interviews, and written assessments, which are further described later.

I envisioned student writing as an integral part of each class period, consisting of a combination of a learning log and a response journal. Students would write about what they were learning in math and I would respond to their jottings. Given the limitations on my time and the students' time, I intended to assess students both midway through the class and at the end of the nine weeks of classes by means of an interview and a written questionnaire and use these "pre-" and "post-" assessments to obtain both "hard" and "soft" data to learn more about what the students were thinking about math.

Journal Writing

My desire was to encourage students to begin to use the language of math, as well as to write more extensively. The students were asked to respond in their journal during the last 10 minutes of class each day to three questions:

1. *What "struck" you?* Was there anything during the session that made you say "aha"?
2. *What didn't you understand about today's lesson?* Was there anything that needed further clarification, more examples, more time?

3. *Did you learn anything new today?* Or did you remember something that had been long forgotten?

Each journal session involved the students in responding not only to the three questions just listed but to a personal question about their previous entry. Because I reacted to students' entries on a daily basis, I also asked them a question related to their prior journal entry, for example, to clarify a comment or to expand on their feelings.

At the onset of the course, I stated that all journal writing entries would be confidential, although I did share with them that this was a research project, and I received their permission to use their writings in the context of my research "story." Students were free to write about whatever they felt compelled to, but I did try to guide them using the three questions. Students were told up front that I did not care about the mechanics of writing. I wanted them to express on paper what they were learning and what issues they had. During the entire process of journal writing, no students ever asked me for help in writing, nor did anyone feel compelled to resort to the dictionary.

Interviews

One-on-one interviews were to take place half-way through the course and at the conclusion of the entire nine weeks. Because of time constraints, the interviews scheduled for midway through the course never transpired. I did succeed, however, in doing the final post-interviewing.

The purpose of the interviews was to elicit input about the value (or lack thereof) of journal writing in the context of a math course. I tried to structure the questions so students would feel free to offer what might be construed as "negative" feedback. I wanted students to be completely honest with me. I was concerned that, because I knew they respected me, they might be reluctant to give me any "bad news" about the class. In the post-interviews, I asked several general questions about the class and included four specific questions about writing issues:

- What did you think about having to write in a journal during class?
- What were the positive and/or negative aspects of journal writing?
- Would you suggest journal writing be included in another math class, or could the time have been better spent? Why?
- Do you ever intend to write in a journal again? Why?

Pre- and Post-Assessment

The written assessment was designed to enable comparison of student's responses at two time points, at the beginning and at the conclusion of the course. The questions were framed to explore change in three general areas that the whole course (not necessarily only the journal writing) aimed to impact:

- *Attitude*—By participating in the math class, would a student's attitude toward math change?
- *Behavior*—Would the math class have an impact on what students do regarding math? Would students be more willing to participate in activities involving math? Would they feel more prepared to take responsibility for math-related tasks?
- *Knowledge*—Although it would be reassuring to learn that students had a more positive attitude toward and were willing to use math, I also wanted to know if they had learned anything new in math. Had the students' knowledge base increased since participating in the math course?

Because I wanted both statistical data and qualitative or anecdotal information, the assessment contained two types of questions (see the Appendix). Part I included 12 statements, involved a rating scale of 1-10, and explored attitudes toward math. Part II included nine open-ended statements designed to provide information on the same three areas—attitude, behavior, and knowledge—with one additional question focused specifically on writing issues.

RESULTS: CONTENT OF JOURNALS

My classroom research experience suggested that journal writing leads to four important outcomes, which are listed here and further discussed later.

- Increases use of the math language
- Allows ongoing assessment of student learning and attitudes
- Gives feedback on classroom techniques
- Provides feedback on the application of math in daily life

Overall, I found the journal writing process to be valuable for both the students and for me as an instructor. Having the opportunity to read what students wrote about math gave me insight into their thinking.

Using the Language of Math

Students may not have written as much as I would have liked (or expected), but they all wrote and some became more willing to write as the weeks progressed. The quality of their writing did not necessarily improve, but the quantity definitely did, and they all expressed their views about learning math. I was pleased to see that they began to use the language of math with more regularity and were even simplifying the vocabulary to suit their needs.

For example, this was the first time I had seen standard deviation defined without having to refer to the unwieldy formula!

> SD is Standard Deviation. It tells me the boundaries (range) of quality . . . how each product differs from each other . . . the margin of error . . . the margin of perfection.
>
> By plotting your findings on [a] histogram, you can get a clear picture of how each product deviates from each other . . . what total is within the 68%-95%-99% range.

The student who entered the following journal jotting now has a sense for metric measurements of length. I expect he will remember these estimates because he has written about them.

> I like how the conversions set up in a ratio because it's getting back to algebra for solving for X (i.e., X could [be] liters, grams, etc.). *(Donna: I love solving for "X" also.)* I can relate to that much easier. Also it's a nice way to check if you are correct in doing your problem. Also relating 1 cm to 1/2 an inch, 1 meter is about 1 yard helps [you] to know if you are in the ballpark with your answers.

The individual who jotted the next note did not enjoy writing. This entry showed me that she was beginning to expand and give specific examples. In earlier journal notes, she would have ended with ". . . figuring out percent from data given."

> Worked with probability today. Had fun figuring out outcomes of different numbers. (Donna: What was fun about it?) My new learning was figuring out percent from data given such as
>
> $$\frac{6}{25} = \frac{24}{100} = 24\%$$
>
> 25 into 100 is 4 and 4 x 6 = 24. Thanks. *(Donna: You're welcome!)*

Ongoing Assessment of Student Learning and Attitudes

Not only did the journal writing offer a glimpse of how the students used the language of math, it became a tool for assessment. When students provided examples such as those presented next, I was able to determine what the students were taking in during the class sessions. In the next journal jotting, the student added arrows to explain how to set up a proportion. I had not used any arrows in my discussion.

> On [today's] lesson
> Correct way of setting up a problem.
> For example:
>
> 5 ml = 3 Tsp
> 10 ml = X Tsp
> (Donna: adding this helps to show even more clearly how all 4 pieces relate)

The following student gave me an indication that she was now more comfortable with the relationship among inches, feet, and yards. She used the journal as an opportunity to check herself on her new knowledge.

> Today I realized that I had a hard time seeing yards (compared to inches and [feet]). After doing a few examples it really [helped] me a lot. I had to draw a ruler on the top of my paper to help me see how to convert inch into [feet] and [feet] into yards.
>
> This helped me a lot to convert what was given. That is why I had no problem with the examples.

12 inches	24 inches	36 inches
1 ft	2 ft	3 ft
		1 yd

> (Donna: Thanks for the visual—a good idea for me to use when I teach this next time.)

I believe journal jottings can be valuable tools for instructors. Comments such as this one gave me insight into how the student makes connections to the real world as well as the value he places on education.

> Answers: I feel more comfortable with math. I have had a lot of conversations with friend(s) and family about the classes I have been

taking. Education is like a power given and your are able to use that power to your advantage like figuring out how much gravel you are actually getting! (*Donna: I agree! Have you had any opportunities to use your new power?*)

On many occasions students wrote that math was challenging. This recurring attitude was true for all students participating in the class. As an instructor this sense of challenge was reassuring. Students seemed to feel that math was not something negative, more it was seen as something that they could rise up to and meet without fear. I suspect, however, that the term *challenge* had a different connotation for each student, and individuals looked at each topic as a different form of challenge.

Math is like putting a puzzle together. You have to ask yourself question[s] about what the end product will be. Having to challenge [your]self to get the end product always excites me. (*Donna: How do you challenge yourself?*) Being determined to find the answer, by knowing what the number[s] are trying to tell us. (*Donna: I like your analogy of math=puzzle.*) What made me realize I loved math was just seeing the math problems again.

On one occasion during the project, students were asked to write a five-line poem about math. (The first and last lines consisted simply of the word "math." The second line consisted of two adjectives describing math, the third line three verbs, and the fourth line was a sentence about math.) The following are examples that reveal students' thinking about math as a challenge.

Math
Sweaty palms/dry throat
Subtract—divide—addition
I find math challenging
Math
(*Donna: Why does math elicit sweaty palms and dry throat?*)

Math
Challenging, positive
Thinking, writing, talking
I really enjoy working with math.
Math
(*Donna: How do you consider math "positive"?*)

Math
Love, challenge
[Hiding], finding, solving
Math is like a game of hide and seek
Math

Feedback on Classroom Techniques

Using manipulatives with adults has been a topic of conversation with many of my fellow math team members. Comments made by students in their journal jottings confirmed for me that they do enjoy using manipulatives, and I definitely intend to continue to incorporate them into any math class I teach—at whatever level of instruction.

What kind of manipulative did I use? I used quite a few: We folded and tore cut-out triangles to prove there are 180 degrees in a triangle. We measured different size circles to determine that the circumference was always a little bit larger than three times the diameter (pi). We used dice when first learning about probability and later in the discussions on theoretical versus experimental probability. (This was a nice lead into sampling processes used at the work site.) We counted the number of candies in mini-bags of M&Ms to create histograms, and we used the same bags to talk about sampling and the probability of choosing various colors.

> It is certainly a better way to learn when we are part of an activity. Experiencing something is better understanding something.

> Today we worked with probability: tossing coins, dice, cards. Just had fun figuring the outcomes out. (*Donna: What was fun about it?*) Did learn that there is a difference when you use the words (or) and (and): and—you will multiply across or—you will add across.

On several occasions, students commented on having "fun." I think the manipulatives helped to bring fun into the classroom. I also believe that learning is facilitated when fun is involved, but that is a question for yet another research project.

> Learning today was fun! You made probability friendly with using the chart and dice with finding even numbers and odd numbers. (*Donna: What do you think of probability now? Has your perception changed? If so, how?*)

Feedback on Application to Daily Life

Whenever possible, I tried to connect math to daily life as well as to the work environment. It was refreshing to see that students were also making the connections.

> "What Struck Me"
> How we use ratio and proportion a lot for everyday life like [your] pay etc. (*Donna: Can you think of examples? Will knowing how to use ratio and proportion change the way you tackle problems?*)
>
> As far as today's class, could we use probability as a way of describing the number of people [that] were laid off? I remember doing this in BPST. Numbers of failures of a disc or tape and why they failed. (*Donna: Could you tell me more about how you did this?*)

Not only did students think about applying concepts in areas outside the classroom, they communicated their learning to others.

> Oh! I am learning. (Donna: Great!) There's no doubt about it. I still want to bring Lou to this class . . . he probably would think twice about parting with his $$$. (*Donna: you can bring your new learning to him!*)

An unanticipated outcome from the content of the journals was the connection to the family. Several times throughout the course, some students mentioned that they were trying various activities at home with their children. Not only were they attempting activities (especially those in which manipulatives were used), but they mentioned that they felt more confident in helping their children with math homework.

I find this particularly interesting because adult educators tend to create artificial distinctions between different types of adult literacy: workplace versus family versus GED versus ABE. The following examples show how students in the workplace have made a connection between "workplace education" and "family literacy."

> Diary for Math...[date]
> I am starting to feel [a bit] more comfortable w/math because working with my [son's] math homework (*Donna: What kind of math is he doing?*) I was amazed and excited that I understood what the question(s) was asking and more important I knew how to approach the problem. (*Donna: It does feel great when you know you're on the right track, doesn't it?*)

In my response to this journal jotting, I encouraged the student to discuss in more detail the kind of math he was teaching his son. He replied:

Finding volume areas of rectangles:

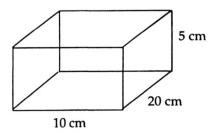

Find the volume of shaded area.

(*Donna: Thanks for the visual! You ought to be comfortable with the geometry section of Prep Math.*)

Finally, the next student articulated one of the real values of communicating mathematically: when we talk about what we learn, when we share with others, we ourselves gain a deeper understanding.

> I feel I now have the knowledge to explain an angle to my children—for me that's important. If I can explain something, I know I understand it. In math [that's] important to me. (*Donna: Why is it important in math for you?*)

RESULTS: INTERVIEWS

A most enjoyable aspect of this research project was the opportunity to interview each student. Interviews were a part of my project because I wanted to find out what students thought about the process of writing.

The students all agreed that journal writing was a new experience for them, not one that they at first readily accepted. One student stated, "At first it felt awkward because I had never done it before." When asked if he would recommend journal writing again, he replied, "ANY class should have journal writing. I feel people will not say anything out loud in a group. I see people all the time be afraid to say how they feel. This way they can."

Another student responded that the process of writing in a journal was "quite different." She stated, "I'm a phone person," and added that journal writing was a good experience because "It is good to write down what is new. You forget new words learned during the day." She felt, however, that it would have been more effective to write down things all during the class time because "I forget what I want to say. I forget at the end."

The following student's comment reminded me of why I began this research project, that is, the mistaken belief held by whole language experts. When asked how she felt about journal writing in a math class, she stated, "I usually can express myself better on paper, but I found it strange to have to talk about math. I had difficulty writing about math." When asked to further elaborate, she said that she likes to write because she can be creative, but in writing about math, "Well, math is math! Math is finding the answer and that's it. Show me the techniques and I'll solve the problem. What's to write about?"

Fortunately, she went on to say that she has changed her thinking about math and writing about it. She later added, "By writing about it, it helps me to remember." A related comment I found most interesting was this: "It's almost like telling someone a secret without fear of wondering what someone else thinks. You gain a trust through writing about it."

Comments such as this one made me realize that journal writing did serve as a useful avenue for students to express themselves. Based on interview comments, students did feel that writing about what they were learning in a math class was helpful in retaining the information and asking for clarity.

RESULTS: WRITTEN ASSESSMENT

The written assessment was intended to provide me with some general feedback on change in students' attitudes, knowledge, and behavior during the course and did not relate directly to the process of journal writing. Some connections could be noticed between comments made in students' journal entries and outcomes of the assessment.

Students did comment in their journals that they were making connections to everyday life. When asked in items #6 and #8 of the written assessment about using probability and geometry in their daily lives, results showed an increase from the pre- to the post-assessment. I am not sure whether this increase is a result of students' greater understanding of what probability and geometry involve, or a result of greater comfort in becoming involved in tasks dealing with probability and geometry. It is possible that journal writing was a factor in this change, but this cannot be determined with certainty given the research design and data I was able to use in my setting.

In having the student participate in journal jottings, I wanted them to become more comfortable with the language of math. In a comparison of pre- and post-assessment results, students appeared to be much more comfortable with the language of the metric system, as well as trying to solve metric problems (items #5 and #11). Again, it is not clear whether the journal writing itself contributed to the increase in comfort level. I believe, based on students' journal jottings, that writing about the math was one factor in helping students increase their comfort with the language of math.

Several interesting responses were received to the open-ended item in Part II: "I think writing in a journal is ____":

> "... making you think when I don't want to!"
> "... a communication tool for teacher and student."
> "... a good way to express your feelings."

DISCUSSION AND UNFINISHED BUSINESS

How will all this information (including the statistical data not presented here regarding results of pre- and post-comparisons) impact my teaching? I plan to continue journal jottings in math classes, at all levels of math. I feel that the journal was a way for students to express themselves, especially those who were reticent to speak up in class. I also believe that students, in writing about what they were learning, became more fluent in the language of math.

Of course, journal writing is not the only way to bring communication into the math class. Oral discussion and problem solving among students in group activities provided opportunities to develop math skills through communication in my math class. Periodically, students were asked to design their own word problems. (chap. 13 elaborates on this and other issues related to writing in a math classroom.) These strategies can all contribute to achieving the goals of the "Mathematics as Communication" *Standard* presented in the Introduction section, which include:

- develop reading, writing, listening, and speaking skills to communicate mathematically
- discuss and make convincing arguments about mathematical problems and reflect on one's own thought processes
- recognize situations involving mathematics, whether it be at work, at home, or in a test situation
- appreciate how the language of math is valuable

Although the results of this research project confirmed for me that writing does have a place in the math classroom, given the many constraints on the research design and assessments I could implement, I am left with several questions regarding journal writing:

1. How can I encourage students to write more? Even though I phrased my responses and questions (to students' journal jottings) to encourage students to explain their ideas more fully, I still had some who would write as little as possible.
2. How do I make sure students are not just writing what they think I want to hear? I had established excellent rapport with my students; many I knew already from earlier classes and from visits at the worksite. I wonder if, because they did not want to "offend" me, they sometimes did not share any "negative" feedback about class.
3. How can I tell what specific activity helped students gain confidence and learn more effectively: manipulatives, style of presentation, journal writing, oral discussions? Obviously, this question begs for more research projects.
4. How do I help the student who is afraid of being wrong? This question, which touches on a broad issue that concerns me as a math teacher, first came up after reading the following journal entry of one individual student:

> New learning is great, especially when you understand it and are able to express it in your own word/thoughts and you're right! (*Donna: Do you feel that you understand whenever you get the right answers?*). Converting numbers is fun when I get the right answer. (*Donna: Could it be fun even when you don't get the right answer?*) Also the process—understanding the process is important—the joy one gets when actually understand[ing] something or anything for that matter. (*Donna: I agree!*) The true joy comes when you actually are able to use what you learn. (*Donna: This is wonderful, and very true!*)

I became concerned after reading the student's answer to the question (about having fun even when being wrong) I asked him regarding his journal entry:

> Can it be fun when I don't get the right answer? Truthfully no! Learning the process and doing the process can be fun but [the] end result has to be [the] right answer in order for me to feel comfortable with myself. (*Donna: Thanks for your response.*)

Similarly troublesome for me as an instructor trying to instill joy in learning math was another written response from the same student, in response to my question regarding the content of his poem on math:

> Why uncomfortable? . . . apprehensive, not picking the right number. You have to be right on target—being close is not good enough. (*Donna: "Being close" is a good start.*)

IMPLICATIONS

My purpose in writing this chapter was to share the results of my experience as a teacher-as-researcher trying to answer the question "Does writing about what is being learned in a math class lead students to a better understanding of and broader insight into math concepts?" Overall, my answer would have to be *yes*. Based on comments made by students and my own experience on this project, I firmly believe journal jotting does have a positive effect on students who are learning mathematics. Learning is enhanced when students make connections to their daily lives. Although adult educators have been strong advocates of this principle, it was reassuring to see my students, in their journal jottings, validate this belief.

In the workplace and in real life, adults combine a variety of skills to accomplish tasks. So, too, should the case be in an adult education class. Rather than isolating math from writing or reading or any other content area, activities should be interrelated to effectively facilitate learning and allow students to use their strengths in one area to assist them in learning in areas where they may be less strong, and express their feelings about mathematical situations and about what they are learning.

Based on this project, I believe there are implications specific to adult education classes, whether in the context of a math class or an integrated ABE classroom.

- Teachers should bring writing into the math curriculum. Students, in their writing, did begin to write in mathematical language. An added benefit of writing in math class is that it provides opportunities for students to use another mode for learning math. Some adults cannot readily grasp math concepts through the traditional math presentation.
- Teachers should use a variety of strategies to teach math. Students responded positively to the use of manipulatives in math class. They shared their enjoyment at learning math con-

cepts using a variety of hands-on activities—dice, paper cut-outs, string, anything to help students visualize and "touch" math.

• Teachers should use journal writing as a method to evaluate teacher instruction. Reading journal entries allowed me as an instructor to assess my presentation of new material. It also gave me an indication of where I needed to provide further support.

Each of the implications could inform further teacher-as-research projects. We are just beginning to look closely at adults and how they learn. Using the *Massachusetts ABE Math Standards* (Leonelli, Merson, Schmitt, & Schwendeman, 1994) or other evolving frameworks as a foundation, further research would give support to tentative conclusions reached in this and in related research projects.

APPENDIX: PREP MATH—WRITTEN MATH ASSESSMENT

Name _____ Date _____

Part I: Rate each of the following statements (circle a number)

1. I become anxious whenever I have to work a problem involving math.

1	2	3	4	5	6	7	8	9	10
Never				Sometimes					Always

2. My stomach churns whenever I am asked to participate in an SGIA*.

1	2	3	4	5	6	7	8	9	10
Never				Sometimes					Always

3. I have thought about taking a college level math class.

1	2	3	4	5	6	7	8	9	10
Never				Sometimes					Always

4. When I read, I tend to skip over any charts or graphs in the text of the material.

1	2	3	4	5	6	7	8	9	10
Never				Sometimes					Always

5. I am comfortable when people around me are talking about the metric system.

1	2	3	4	5	6	7	8	9	10
Never				Sometimes					Always

6. I use geometry when I'm working on projects at work.

1	2	3	4	5	6	7	8	9	10
Never				Sometimes					Always

7. I use geometry when I'm working on projects at home.

1	2	3	4	5	6	7	8	9	10
Never				Sometimes					Always

8. Probability is useful in my daily life, both at home and at work.
       1   2   3   4   5   6   7   8   9   10
       Never   Sometimes   Always

9. When I go shopping, I try to figure out in my head how much I save on sale items.
       1   2   3   4   5   6   7   8   9   10
       Never   Sometimes   Always

10. I am willing to keep track of the household budget for my family.
       1   2   3   4   5   6   7   8   9   10
       Never   Sometimes   Always

11. I am comfortable solving problems involving the metric system.
       1   2   3   4   5   6   7   8   9   10
       Never   Sometimes   Always

12. Fractions "intimidate" me.
       1   2   3   4   5   6   7   8   9   10
       Never   Sometimes   Always

Part II: Complete each sentence in your own words.

A. When I hear the word "math", I _____.

B. I think probability is most useful in _____.

C. When I see a problem involving a fraction, I _____.

D. If you asked me what 25% of a number is, I would _____.

E. To me, "volume" means _____.

F. I use a calculator whenever I have to _____.

G. When I hear someone talk about the "average" of something, I think
_____.

H. I think writing in a journal is _____.

I. What I hope to get out of this Prep Math class is _____.

*SGIA is Small Group Improvement Activity, the company's concept of Team Activities.

REFERENCES

Leonelli, E., Merson, M. W., Schmitt, M. J., & Schwendeman, R. (Eds.). (1994). *The Massachusetts ABE math standards project* (2 vols.). Holyoke, MA: Holyoke Community College.

National Council of Teachers of Mathematics (NCTM). (1989). *Curriculum and evaluation standards for school mathematics.* Reston, VA: Author.

13

The Challenge of Diversity in Adult Numeracy Instruction

Harriet Hartman
Rowan University

This chapter synthesizes and reflects on some of the theoretical and practical perspectives presented in this volume and makes more explicit some of the implications of diversity as a challenge to teaching adult learners, particularly in relation to numeracy. The chapter reflects on the diversity deriving from gender, race and ethnicity, immigrant status, social class, and learning disabilities. This is not meant to be an exhaustive coverage of diversity, but rather a stimulant to thinking about the challenge of incorporating the needs and contributions of every adult learner (and, presumably, younger learners as well).

INTRODUCTION

Several chapters in this book point out ways in which adult learning of numeracy presents special circumstances and challenges. Adults' motivations for learning are based on their perceived needs and heavily influenced by their life-centered orientation to learning. Their life experience is their greatest resource for learning, but it also presents the greatest diversity, as adults come to their classes with divergent needs, goals, backgrounds, and capabilities (see Foerch, chap. 6, this volume). Although adults are generally self-directing, previous learning experiences may initially prompt them to regress to dependent learning styles, particularly when it comes to math learning (see Coben, chap. 2, this volume). Dependency, however, conflicts with their general way of functioning as adults and is an obstacle to be overcome. Adults' previous experiences with learning math may have alienated them from seeing the contribution of increased numeracy skills in their everyday lives. At the same time, adults' needs for self-validation require a learning environment in which their own life experiences can be appreciated and harnessed for increased self-growth and development (Knowles, 1990), including but not limited to the area of numeracy.

Several issues are common to working with most diverse populations:

- The challenge of empowering adult learners and raising their self-esteem so that they use their own experience as a foundation for learning
- The need to overcome past legacies of negative teacher or student attitudes in learning situations
- The challenge of promoting and accepting achievement and further learning as positive and relevant goals, after these have been rejected in the past as legitimate or accessible pursuits
- Differential costs of learning
- The difficulty of crossing "learning boundaries" between everyday experiences or tasks and the classroom
- Learners' differential abilities to articulate needs
- Learners' differential access to the abstract reasoning skills needed for higher math.

These issues are discussed here in relation to different facets of diversity, mainly gender, ethnicity-race-immigrant status, and social class. Some ideas are offered about how learning contexts and experiences can build on the diversity of different populations and thus strengthen the resulting learning. Subsequent sections address in more detail adults with learning disabilities, women and math, and, more generally, a constructivist approach to learning.

Gender

Many of the adult learners are women. Some have come to class come to better their employment opportunities, some to enhance their parenting skills, some to increase their working knowledge of mathematics. Extensive research has established that: (a) the learning context differs for girls and boys (e.g., American Association of University Women, 1992); (b) this difference extends to a "chilly climate" for the women in advanced educational contexts (e.g., Crawford & MacLeod, 1990) and in adult basic education (McLaren, 1981); and (c) some differences in the learning context are most pronounced in the mathematics and science subject areas (Baker & Jones, 1992).

Women (and girls) are disadvantaged in that their interests are often neglected or omitted from mathematical subject matter. When typically female activities are noted, they are often isolated rather than integrated into the wider curriculum; women often have less access to mathematical instruction and to extracurricular mathematical activities. Each of these is likely to contribute to alienating women of all ages from mathematical subject matter, thereby reducing their identification with the subject and the skills it requires and generally reducing their motivation for pursuing the learning of mathematics.

School counselors and teachers may further discourage girls' motivation by steering them away from occupations needing mathematical or scientific training, in which women are traditionally underrepresented. As a result, women's understanding of the importance of knowing mathematics may be hampered, and their expectations for performing well in mathematics are reduced. As Muchinsky (1990) suggests, expectations about the likelihood that investment in effort (and learning) will lead to desired outcomes affect motivation to work. Girls' lowered expectations would therefore further reduce their inclination to invest effort in mathematics.

More generally, young women are more prone to follow a pattern of "learned helplessness" in school, a loss of self-confidence, and may often experience sexual harassment. They may consider men's "claims to knowledge" as superior to their own, based in part on their own daily and family experiences, further alienating them from school-based learning (Luttrell, 1992). As a result, even talented girls may shy away from further academic achievement. Although awareness of these gender biases has been raised, gender disadvantage still abounds and results in schools "shortchanging" girls (American Association of University Women, 1992).

Teachers of adults must be aware of the gendered context of previous learning and the possibility that it may present a hidden obstacle in the current learning experience; educators should experiment with ways to overcome the alienation women may carry with them from the past. Because of previous biases they have encountered in the curriculum, women may need extra reinforcement to bridge the gap between the present learning situation and their prior experiences and to be willing to apply the present learning to their nonclassroom activities.

Harris' chapter, which follows, suggests ways to integrate traditionally female activities such as needlework into the mathematics curriculum, validating such activities as mathematical. It also suggests a way of bridging between what is often considered nonacademic activity, alienated from the learning context, and classroom activity. Such activities are necessary to reduce gender bias in the curriculum and allow for a wider basis for identifying with the subject matter, at the adult level as well as at every level in the learning process.

Social Class

Research on the "hidden curriculum"—the ideological messages accompanying the learning experience both within and outside the classroom—suggests that the learning context differs for lower class, middle class, and upper class students. In a disturbing ethnography of several schools in neighborhoods that differ by class, Anyon (1980) demonstrated that children from more privileged backgrounds are more likely to be exposed to rich, sophisticated, and complex material than are working-class students; further, their independent reasoning and problem-solving abilities are more likely to be developed as part of the grooming for the types of occupations their parents and teachers expect them to enter. These students' innovative approaches and independent interests are more likely to receive validation in the classroom, reinforcing their self-esteem and providing a much smoother bridge between their outside lives and classroom activities. These students are empowered by the respect their teachers show them and the messages implicit in their learning environments: each of these student's ideas are important and worthwhile, not only in the classroom but in the wider world as well.

Middle-class and especially working-class students have a more difficult time crossing the boundaries between school and the "street," often thinking of their own "street-smarts" as alien to the classroom (Luttrell, 1992; Wallace, 1995). Further, they may see school learning as a threat to their working-class culture, which is based on what they consider "common-sense" (Luttrell, 1992), reducing the extent to which they can accept school-based learning as relevant to their life situations. As MacLeod's (1987) seminal ethnography of working-class culture and high school achievement shows, cultural contexts place differential values and valences on academic achievement,and on the types of reasoning conducive to mathematics skills, which may carry over into the adult context of learning numeracy. Educators must be sensitive to these inner conflicts, which may not be articulated but may hinder application or transfer between the numeracy classroom and the (working-class) learner's world. A constructivist approach to learning (see Coben, chap. 2, this volume) may enable the educator and the learner to work together to build a bridge between coexisting knowledge bases.

Race, Ethnicity, Immigration

Relatively little research has been conducted on the correlates of mathematics learning in different racial and ethnic subgroups. Overall, more research has been done on differential achievement, but less on factors that may affect the learning process (including underlying reasoning skills) or on differences in the learning context experienced by such subgroups. Much of what has been done is reviewed in Cocking and Mestre (1988) and Secada (1992), yet virtually all such research relates to school children, not to adults learning mathematics. Thus, little is known about different cultural perspectives on the knowledge and reasoning skills needed for learning mathematics at different levels, perceived conflicts between a subculture and academic achievement in mathematics (much as the working class culture may perceive a conflict, as noted earlier), or the role of linguistic diversity. Bernstein's (1977) work on "elaborated" and "restricted" language codes suggests that students from certain linguistic contexts may develop through their use of language a greater proclivity to abstract and logical reasoning.

These and related issues may be especially important when we consider the teaching and learning of mathematics in immigrant populations, which in many developed countries constitute a major client group for adult education programs. Immigrants face challenges of adaptation to new patterns of everyday life, rebuilding personal networks and learning the host country's cultural codes and behavioral rules. These demands on immigrants may influence the amount of energy and attention the immigrant adult learner can give to numeracy (and literacy) issues. The immigration experience may also damage—however temporarily—the adult's self-esteem and self-confidence, which may affect both the ability to learn and the motivation to work hard. Further, language limitations may make it difficult for the immigrant (or bilingual) learner to discern between a math principle or concept that is not understood due to a semantic difficulty, which a quick translation could overcome, and more general problems having to do with "real" knowledge gaps in mathematics. It becomes important to give the immigrant learner a chance to voice concerns and difficulties and connect between the new material being presented and prior knowledge (perhaps known or communicable only in another language). Making the material being learned more relevant to the learner's own life context may help overcome what at first may seem like an insurmountable boundary.

Learning Disabilities and Handicaps

Chapter 9 by Sacks and Cebulla (this volume) discusses some of the issues involved in assessing learning difficulties and dealing with students with different problems in the classroom. Educators should also be aware of the impact of earlier "special education" experiences and special "tracking" that may have separated many of these learners from "normal" students and colored some of these students' perceptions of the learning context and their place in it. Oakes (1985, 1989) has shown that lower track students are taught at a slower pace and presented with less challenging material, under teachers who expect less achievement and often are less experienced (summarized in Sedaca, 1992).

One of the consequences is that students placed into "lower" or nonacademic tracks learn to see themselves as poorer achievers; students in adult numeracy classes who were previously placed in "special education" may thus see themselves as not capable of learning math or understanding abstract material. These adults, who are trying to make up for gaps in their earlier education, may be handicapped by an alienation from new learning processes due to earlier educational experiences. Although as adults they are generally better able to articulate their needs and less likely to suffer from embarrassment, their earlier experiences may be obstacles in reaching achievements that are actually well within their reach.

SUMMARY AND IMPLICATIONS

Each adult learner carries "baggage"—a set of previous learning experiences and culturally based attitudes toward schooling, learning, mathematics, and the skills needed to become numerate. Often this "baggage" serves as a barrier to full participation in and benefiting from the learning situation. By understanding the cultural, social, and educational contexts that have influenced the student, adult educators may be better able to help students utilize their life experiences in the learning context, reinforcing self-esteem and validating alternative approaches to problem solving. A practical and efficient way of understanding students and their needs may be through a constructivist approach, building knowledge with rather than for students, facilitating the articulation of students' needs and starting points, and staying in touch with the learning process as it unfolds.

Two chapters in this volume, by Sacks and Cebulla and by Harris, suggest ways to apply these insights to different populations and situations of adult learners. They show why it is important to have learning environments that take into account students' experiences, motives, goals, styles, and continuous reactions to the learning process. Without such an environment, the students may not be able to build connections between their wealth of experience and the world of numeracy, limiting the applicability of what they learn to the(ir) world outside the classroom.

Sacks and Cebulla show us how developmental disabilities are often camouflaged and misdiagnosed. The true nature of the "learning boundary" that may have prevented "slow" learners from fully benefiting from formal learning experiences in the past is often misunderstood. By harnessing students' own insights about their learning difficulties and reasoning powers, and validating their "natural" thinking processes, teachers are able to meet students where they are and help them break down the barriers to learning.

Harris' chapter (this volume) focuses on women who do traditional craft activities in developing as well as developed countries. She shows how unlikely elements of such women's life and work experiences and their unique mathematical skills can be engaged to support or enrich the numeracy learning situation. Awareness of the value of these women's own resources for numerate behavior can bridge the gap between the passive dependency and silence that these students expect to experience in the math classroom and the self-directedness they bring to much of the rest of their life. Harris also claims that, although math education purports to be value-free, it often is, in fact, laden with overtones that may alienate students who come from nonmainstream backgrounds and modes of operation. She urges us to use a constructivist approach to learning and start instruction from the students' own experiences, so as to bridge the gap between students' traditional craft activities and conventional mathematics.

In various other chapters of this volume, strategies for such "constructivist" learning are suggested, with the goals of encouraging students to actively engage in learning constructed mutually by student and teacher. Van Groenestijn (chap. 15) develops the supermarket shopping strategy to enable assessment of students' mathematical knowledge and reasoning skills in the context of their everyday experiences. The process writing introduced by Hicks and Wadlington (chap. 10), and approaches described in other chapters in Section 3, engage students in goal-setting, reaction, and evaluation, and enable them to jointly develop literacy and numeracy skills. Clemen and Gregory (chap. 4) suggest how to use the needs perceived by all adults for efficient problem solving as resources for connecting numeracy learning to real life choices and decision making.

When working with adult learners, it is easy to fall into a traditional math learning pattern that can block students from applying numeracy in their daily lives. It is a challenge to implement constructivist principles in the numeracy learning context, when working with populations whose life experiences have so often been excluded from the classroom learning situation. The chapters in this section should be understood as examples and inspirations for such an implementation. When the adults' own experience can be harnessed, the learning experience is likely to be enhanced, exciting, appropriate, and rewarding to both students and teachers.

REFERENCES

American Association of University Women. (1992). *How schools shortchange girls.* Wellesley, MA: College Center for Research on Women, American Association of University Women, Educational Foundation.

Anyon, J. (1980). Social class and the hidden curriculum of work. *Journal of Education, 162,* 67-92.

Baker, D., & Jones, D. (1992). Opportunity and performance: A sociological explanation for gender differences in academic mathematics. In J. Wrigley (Ed.), *Education and gender equality* (pp. 193-203). Washington, DC: Falmer Press.

Bernstein, B. (1977). Social class, language, and socialization. In J. Karabel & A. H. Halsey (Eds.), *Power and ideology in education* (pp. 473-486). New York: Oxford University Press.

Cocking, R., & Mestre, J. (Eds.). (1988). *Linguistic and cultural influences on learning mathematics.* Hillsdale, NJ: Erlbaum.

Crawford, M., & MacLeod, M. (1990). Gender in the college classroom: An assessment of the "chilly climate" for women. *Sex Roles, 23,* 3-4, 101-22.

Knowles, M. (1990). *The adult learner: A neglected species* (4th ed.). Houston, TX: Gulf Publishing.

Luttrell, W. (1992). Working-class women's ways of knowing: Effects of gender, race, and class. In J. Wrigley (Ed.), *Education and gender equality* (pp. 173-191). Washington, DC: Falmer Press.

MacLeod, J. (1987), *Ain't no makin' it*. Boulder, CO: Westview Press.

McLaren, A. T. (1981). Women in adult education: The neglected majority. *International Journal of Women's Studies, 4*(2), 245-258.

Muchinsky, P. (1990). *Psychology applied to work*. Belmont, CA: Brooks/Cole Publishers.

Oakes, J. (1985). *Keeping track: How schools structure inequality*. New Haven, CT: Yale University Press.

Oakes, J. (1989). Tracking in secondary schools: A contextual perspective. In R. E. Slavin (Ed.), *School and classroom organization* (pp. 173-175). Hillsdale, NJ: Erlbaum.

Secada, W. (1992). Race, ethnicity, social class, language and achievement in mathematics. In D. Grouws (Ed.), *Handbook of research on mathematics teaching and learning* (pp. 623-660). New York: Macmillan.

Wallace, N. (1995). *From the street to the classroom: Bridging the numeracy gap*. Unpublished manuscript.

14

Mathematics and the Traditional Work of Women

Mary Harris
Institute of Education,
University of London

This chapter looks at several common and widespread work activities in which women have traditionally been involved. Such activities are rich in mathematics, yet have paradoxically been ignored or trivialized by the mathematics education community. In the first half of the chapter, three examples of such activities are presented, the rich mathematics in them is analyzed in detail, and classroom activities building on such experiences are illustrated. The second half of the chapter raises and attempts to answer three questions: Why does this mathematics not feature in mainstream mathematics education?, How can it be made to feature?, and Why should it be made to feature? In answering these questions, political and historical processes involved in the gendering and devaluation of mathematics education for women are explored. Suggestions are made to reevaluate the content of mathematics education and introduce changes that can raise women's confidence and involvement in mathematics. The chapter aims to raise awareness of current problematic practices in mathematics education, as well as to suggest practical steps that educators can undertake in this regard.

INTRODUCTION

In order to introduce the topic of mathematics inherent in the traditional work of women and to examine the educational promise inherent in women's work experiences, the nature of traditional methods of research into the mathematics people use at work is first explored. This may seem to be detour, given the "political" agenda of the chapter as presented earlier, but it is essential for understanding the (limited) nature of what is known about the mathematics inherent in people's work.

Research into the mathematics that people use at work commonly includes questionnaire and interview techniques in which researchers elicit responses from workers to questions about specific mathematical skills they use. For example, a researcher interested in the *frequency* with which particular skills are used, may read questions from a prepared questionnaire, such as "How often in the course of your work do you add fractions?" (Harris, 1991a). Alternatively, as was done in the research leading to the influential Cockcroft Report in the United Kingdom (Cockcroft, 1982), experienced mathematics education practitioners used a less formal approach and talked generally with workers about their schooling and training before asking for examples of their use of mathematics at work. Here the research interest was in *identifying which* mathematics skills were used at work. A major disadvantage of these research methods is that workers often fail to recognize or credit themselves with the mathematics involved in the work they are doing, either denying it to the researchers or dismissing it as common sense (Harris, 1991a). The result is an inaccurate account of mathematical activity in work situations.

More recent research is concerned with mathematics activity within the total context of what workers do, attempting to assess the effects of the workplace context on how people use mathematics. Extended interviews and comparative testing as used by Carraher and others (Carraher, 1991) provide detail of people using very different arithmetic techniques when dealing with situations that embody similar mathematical principles, according to whether they are in educational or workplace settings. Participatory observation studies by Lave and others (e.g., Lave, Murtaugh, & De la Rocha, 1984) result in a detailed account of high accuracy in arithmetic when it is embedded within the daily business of supermarket shopping, compared with similar arithmetic done with the pencil and paper of educational settings. Scribner (1984) used ethnographic methods to study the mathematical behavior of workers in a milk-processing plant and revealed problem-solving strategies that could not have been revealed by responses to an arithmetic skills questionnaire.

In summary, what is really known of the nature and extent of the mathematics that people do when they go about their daily work is very much a feature of the research orientation used to elicit it. Simply asking people what mathematics they do tends to produce negative responses. Research methods that include an involvement of the researcher in the processes of the work under investigation produce a richer and more realistic picture.

Women, Mathematics, and Work

Some recent work of my own with needleworkers details arithmetic, geometry, and problem-solving behaviors of women who sew, knit, and weave in the contexts of both paid and unpaid work. It is from this work that I have synthesized the stories that follow and analyzed their mathematical content. The analyses are followed by examples of classroom activities for mathematics learning, starting with textiles, that have been developed from this work. Throughout the stories I have stressed mathematical terms, or terms with mathematical implications, with italic script. The first story is longer, in order to highlight the prevalence and scope of mathematical elements in women's work situations. The other two stories are shorter and mainly illustrate additional types of mathematics that go beyond what is already described in the first story.

These three stories form an introduction to the discussion of power and gendering of mathematics issues I hinted at earlier and which I discuss in the second half of the chapter, in the context of answering three questions: Why does women's mathematics not feature in mainstream mathematics education? How can it be made to feature? and Why should it be made to feature? I argue that the origin of all three questions lies in social and historical factors that have effectively gendered both mathematics and women's work, defining mathematics as a male subject and needlework as the stereotype of femininity.

The resolution of the three questions lies in a reevaluation of the content of mathematics education and of stereotypical women's work, and in developing the cognitive content of instructional programs for the purpose of raising women's confidence and involvement in mathematics. The Implications section that closes this chapter shows how these ideas can be implemented by describing three educational projects revolving around purposes of advancing cultural confidence, economic development, and the educational enrichment of students.

STORY 1: PLAIN KNITTING

A woman is about to knit a sweater for one of her young grandchildren. She is an experienced knitter and will not be using knitting instructions in the form of a published pattern or recipe. Ask her what mathematics she will be using and she laughs at the thought. She does not use mathematics at all; she always disliked the subject at school and could never do it. She has always managed perfectly well without it.

She has a *large amount* of yarn that she *estimates* is *about enough* for the garment, and a collection of knitting needles of *different thicknesses*. She *measures round* the child's chest, *along the length* of his arm, and the *length* from his shoulder that the sweater is to be, for he has *grown a lot* lately. She uses *inches* instead of *centimeters* because the yarn she will be using is *quite thick* and *centimeters are too small for a meaningful measurement*. To the chest *measurement* she *adds six inches* because the child will be wearing other clothes under the sweater and he needs to be able to move about comfortably. This *additional measurement* is called *the ease*. Because she has *added* it to the *total measurement all round* the child's body, she will use *half of it* when she comes to knit the *two separate pieces of the front and back* of the sweater. There will be a *quarter of this measurement* on *each edge of the front and back*. The knitter has *included an allowance of about half an inch at each edge* for sewing the sweater together. She also knows very well that a child's head is *proportionately larger* than an adult's and she will *allow for* this when she comes to *plan the top* of the sweater. After all, the child needs to be able to put it on and take it off without tearing his ears.

Having *taken the measurements*, the knitter then *needs to work out how many* stitches she will have to use to obtain them. This *depends on the thickness* of the yarn and needles she is using as well as the *tightness* of her own style of knitting. What she actually needs to know is the *number of stitches per inch* she will be working, *a ratio* that in the field of knitting is called *tension* or *gauge*. So she takes some yarn and some knitting needles and casts on *enough stitches to make a row about six inches long*. She then knits *enough rows to form a square*. She casts off and allows the *square* to relax while she makes a cup of coffee. Possibly she may feel that the *fabric* of her sample is a bit *open* or *loose*. This sweater is to be a winter one that needs to keep out the wind, so she will want the fabric to be *fairly tightly constructed*. If her first *sample square* has produced a fabric that is too loose she will have to take a pair of needles of *smaller diameter* and knit another sample *knowing that her gauge will be tighter*. Had she had less experience she might have *experimented* with *different diameter needles* until she was happy with the *tightness* of the fabric, before *working out her gauge*.

From the *measurements* the knitter *has taken* and the *gauge she has calculated*, she can *work out how many stitches* she will need to cast on. She takes a *tape measure* and lays it *across the square, parallel to the cast-on side,* as in Figure 14.1.

From the tape she reads off *how many stitches there are within four inches* and then *divides the number of stitches she has counted by four* to find the *ratio* she needs. She does not *count* the stitches *near the edge of her square* because she knows that the *gauge* there is likely to be *distorted by the edge* of the fabric and *therefore* will not give her *an accurate result.* She then refers back to *the measurement* of the back of the sweater (which is *half the total measurement* of the chest *plus half the ease*). Because she knows *how many stitches per inch* her sample produced, she can now *work out how many stitches to cast on.* She *does the calculation,* casts on, and begins her work.

A few days later she has completed the back and is ready to start on the front. This will be the same as the back up to the *curve* at the neck-line that allows for the child's head to pass through the finished garment.

She takes a comfortable T-shirt the child already has, one that fits well at the neck, and *traces the shape of the curve of the neck* onto a piece of card, in fact, the back of a cereal packet she has flattened out. She then cuts out the *curve* so that she has a *template for its shape.* She places the template on the knitted back that she has already completed and notes *how many rows and stitches* she will have *to get rid of inside the curve* so that she can form a shape that is both *accurate and symmetrical.* She jots down her instructions to herself in her own *shorthand code* on the card itself. By placing the card template so that the *ends of the curve lie on the shoulders of the back,* she can also find out *how long she needs to make the front* before she has to start shaping the curve of the neck (see Figure 14.2). She *measures the length from the bottom of the sweater to the bottom of the curve* and adds the measurement to the notes she has already made on her card. She has

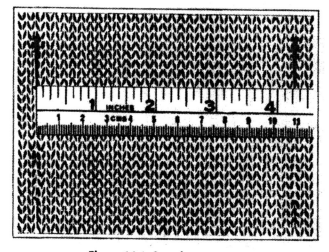

Figure 14.1. Sample square

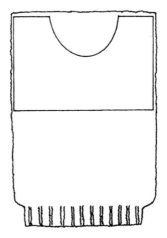

Figure 14.2. The curve of the neck

quite a collection of these cards by now and they make a nice *record of growth* of her different grandchildren.

When she has completed the front, the knitter joins it to the back at the shoulders. She then *estimates how far up* the front a sleeve should start, marking the place on *one edge* of the front with a safety pin. She folds the unfinished garment *in half lengthwise* and *matches the equivalent sleeve point* on the *other edge*. She unfolds it, then refolds the garment *along* the shoulders so that she can place safety pins on the back in the equivalent *position* to those on the front. In this way she ensures that the sleeves will be *symmetrical in both depth and location*. This particular knitter likes to start knitting the sleeves from their *tops*, picking up stitches from *the edges* of the joined front and back. This makes it easier to *add to the length* of the sleeve as the child grows or to unravel the knitting from the cuff to the elbow if the child wears out or damages the elbows or ends of the sleeves before he has outgrown the garment. To ensure that both sleeves are the same in *width, length,* and *shape,* she *calculates* when to cast off at *both ends* of a row, for sleeves *must reduce evenly in width* toward the cuff. She *makes a note of how many stitches to cast off on which rows,* adding it to her other notes on her template.

When the sweater is nearly finished, the interviewer discusses with the knitter what she has been doing and asks her again about the mathematics she has used. Again she laughs, a little impatiently this time. She has already explained that she does not do any mathematics. She has simply been using her common sense, her own way of *saving time* and *not wasting materials*, making sure the sweater *fits first time*.

Mathematical Analysis of Story 1

Whether a sweater fits its intended wearer depends entirely on the principle of ratio, the number of stitches per unit width of knitting. Both imperial and metric units are used by knitters depending on the thickness of the yarn and personal preference. If centimeters are used the division sum involved in the ratio calculation uses larger numbers in the denominator because centimeters are a smaller unit. However, if stitches are counted across 10 centimeters, then the sums are easy. The knitter in this story uses the idea of symmetry in designing the sweater, making the whole garment, and placing and shaping the neck and sleeves, both sides of which must be identical. The context of making the sweater is the problem-solving exercise of ensuring that the garment fits the wearer, using the most economical methods in both time and materials.

A summary of what this knitter has been doing is:

- Measuring in metric or imperial units
- Using the process of trial-and-error estimation and testing
- Rational judgment and problem solving including the choice of appropriate arithmetic for particular circumstances
- Ratio calculation (division)
- Counting
- Estimating position, size, and shape (sleeves)
- Estimating and measuring shapes including squares, rectangles (the back of the sweater), and trapezia (the sleeves)
- Drawing a curve with estimation, measurement, and judgment of symmetry in its depth and shape
- Using the idea of symmetry throughout the exercise
- Taking notes

Suggested Classroom Activity Based on Story 1

There have been objections that not everyone (including many women) knows how to knit. However, inability to knit need not be a reason for not undertaking tasks that arise from it. (Inability to play football is not usually regarded as an obstacle to working problems based on its scoring.) This activity, taken from a pack of teaching materials developed by the writer but now out of print, has been used in schools and in adult education in England and other countries. It has been found to be particularly rich if it is done in groups because this encourages discussion. Asking students to note down any mathematics they use while they work and talk can produce an amount that often surprises them.

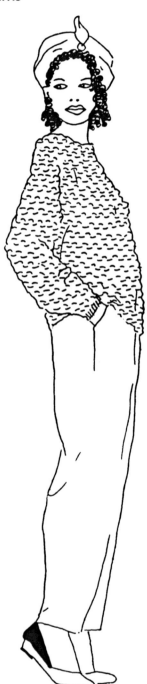

Sweater

I want to make a sweater like this
for myself.

I can make it from rectangles with
some shaping at the neck and the cuffs.

I have knitted a square in the yarn
I have chosen and counted the
stitches and rows in 5cm. My tension is 8
stitches and 14 rows to 5cm.

My own measurements are:
Bust 90cm.
Length 60cm.
Underarm sleeve seam 42cm.

Please,
1. Sketch a design for me.
2. Write in the measurements.
3. Write out the instructions.

MiW Cabbage
Drawing by Akua Bonsu
© Maths in Work

Activity 1: "Sweater." © SMILE Mathematics

However, because students often forget to make such notes once they are involved in the task, it can be useful to leave a taperecorder running so that interactions can be analyzed together once the task is completed.

Experience shows that secondary school pupils and many adults have a tendency to underestimate the size of their heads and to be surprised by its measurements. Primary school children are often used to measuring each other as part of their learning about measurement. Younger or less experienced students also often fail to make allowance for the ease, only discovering it when they make up a paper model of a garment they have planned and find that they cannot move inside it.

A knitter's version of the activity can be made by omitting the tension figures given in Activity 1 and asking students to produce their own. Developments from the initial task can arise if students are asked to design their own garments. A further development possibility arises in adding decoration to the garment (see Story 2).

STORY 2: A FAIR ISLE KNITTER

In the Fair Isles off the coast of Scotland, another knitter sits by the turf fire in her cottage on a long winter evening and plays about with some new Peerie designs (see Figure 14.3) to work into a sweater she will make to sell to summer-time tourists.

Before proceeding, I should note that the words "design" and "pattern" cause endless confusion when writing about knitting. In everyday use, "pattern" is used both for the knitting instructions that knitters buy from woolshops and for any design worked into the fabric of knitting. "Design" is used for the process of creating a new shape and style and for the pattern that decorates the garment. When I try to be consistent by using the word "pattern" for written instructions and "design" for the decoration, I immediately hit problems like the one that follows. Traditionally, the Peerie designs of the Fair Isles, such as the one depicted in Figure 14.3, are called "Peerie Patterns." To call them "Peerie Designs" would fit my attempts to be consistent but would sound strange to a Fair Isle knitter. I have therefore only been consistent up to a point, and having drawn readers' attention to the problem, ask their understanding.

Traditional Fair Isle *designs* are made by taking a *shape*, a *motif*, and *repeating it along a line* of knitting. This knitter likes the designs called *Peerie Patterns* because they are *small* and neat and she can be *inventive* and *systematic* in *putting them together*. "Peerie" is an old Celtic word meaning *little*, and Peerie patterns are by tradition *no more than seven rows deep*; if they were more, then they would no longer be Peerie. Peerie patterns are also *no*

Figure 14.3. Peerie patterns

more than seven stitches wide because of the way the fabric of the garments is knitted. Genuine Fair Isle sweaters are bright with many colors, but it is difficult to keep all the yarns under control if *more than two* colors are used *in any one* row. While knitting a row in *two colors only,* the color not in use lies at the *back* of the work. When it is required by a change in color in the design, it is brought to the *front* of the work, leaving a *loop along the back* behind the stitches just knitted. It is this *double thickness* of knitted fabric and loops with the air trapped between that makes these sweaters so warm. If, however, the *loop at the back of the knitting is more than seven stitches long,* it will hang down inside the garment, not only *distorting the insulating double layer* but also making a trap for the fingers as the wearer puts it on.

The Peerie Pattern designer is, in fact, working on a *matrix of up to 7 x 7 stitches.* She can, of course, use *numbers smaller than seven* and work on a *matrix of 6 x 4 or 7 x 5.* But whichever *numbers* she uses, the *characteristic of the designs* is that they are made from *motifs repeated in strips* across the garment.

The knitter knows that there are different *ways of making a small motif* into a *symmetrical pattern repeated along a line* of knitting. The motif may itself be *symmetrical; that is, when a line is drawn through the middle of the motif, both halves are the same: they would fit over each other* exactly. Such motifs are said to have *mirror symmetry.* Other motifs may have *rotational symmetry; that is, they can be turned until they fit themselves again.* If a motif is *not symmetrical* it can be made so by "flipping" it about an imaginary mirror line before repeating it and its mirror image together as the motif is *translated along the row.*

This knitter likes to *find different ways of combining her motifs.* When she designs, she likes to *invent motifs* on *squared paper* that represents knitted stitches, taking the same *motif* and *turning or flipping* it before repeating it along a line of design.

Making the *patterns from a repeated motif* is only *part of the problem,* however. The knitter needs to *fit her row of motifs on to the sweater so that there is no break in the design round the whole* garment. Traditional Fair Isle sweaters are, in fact, knitted *round and round,* that is, as a *helix.* This makes the knitting easier since it does not need to be turned at the end of each row and, because there are no seams that can be pulled apart, the result is an evenly warm and windproof sweater. But because the sweater *has to fit* someone, it has to be made on a *particular number of stitches calculated by the gauge* used by the knitter. Suppose the knitter worked out that she needed to *use 150 stitches* and her *motif was a repeat of seven stitches. Would the design fit exactly? If not, what would be left over at each side? If she adjusted the number of stitches in the row to fit the motif exactly, by how much would that change the size of the sweater? Would that change be significant? Which designs would fit within 150 stitches—those based on 4? on 5? on 6?*

During the summer, two mathematicians on a holiday walk by and see the finished sweater hanging outside the cottage. They go closer to study its design. One of them buys the sweater to wear in the new term. It will make a good illustration for his undergraduate course in symmetry.

Mathematical Analysis of Story 2

The knitter has been doing what a mathematician might call "symmetry transformations." In mathematics, the symmetry of an object or a motif can be described or quantified by the ways in which it can be made to look like itself. For example, a flower with five petals (and with both sides of each petal the same) has five-fold rotational symmetry. Thus, five turns of 72° each will make it look as it did to start with. One of the transformations that mathematicians talk about is *rotation*, which refers to the way in which an object is turned about some point. The others are *reflection*, as in imaginary mirror line, and *translation*, that is, moving the object along a line or across a plane. Transformations can also be combined. For example, a combination of reflection and translation is called a *glide* (see Figure 14.4); this is the pattern of footprints in the sand.

A "strip" or "frieze" pattern is made by combinations of transformations with translations along a line. Most mathematics textbooks note that there are only seven ways in which a motif can be combined if the strip pattern is to remain on the line. These ways are illustrated in Figure 14.5. The idea of repeating a motif along a line can be extended to repeating it over a surface, in other words, by repeating it in two directions at once. This can only be accomplished in a limited number of ways.

The Fair Isle knitter in the story has been:

- Drawing symmetric and nonsymmetric motifs within a 7 x 7 matrix
- Transforming motifs before translating them
- Creating different patterns from the same motif through different transformations
- Translating motifs along a line
- Calculating the number base for motifs that can fit into a required number (of stitches)

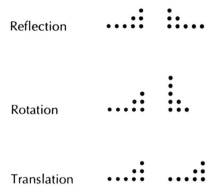

Reflection

Rotation

Translation

Figure 14.4. Reflections, rotation, and translation

Note: The dots represent knitted stiches of one color against a background of another color

Suggested Classroom Activity Based on Story 2

The following activity is published by SMILE, a secondary mathematics program in which teaching materials are designed by practicing mathematics teachers. It has been adapted from the same pack of learning materials as Activity 1 earlier. An example of a similar activity conducted by a student teacher is given in Miller (1991).

STORY 3: PATCHWORKERS

A women's group has met to plan the making of a quilt to be presented at an *auction* intended to *raise funds* for their church. They have collected scraps of cloth from every household in their community and the aim is to display the finished quilt before it is auctioned so that everybody can have the opportunity to identify themselves with it through the recognition of their own contribution. The quilt will be made entirely of *patchwork* of small pieces *all the same shape and size* or possibly of *two different shapes*, sewn together in such a way that *there are no gaps in between the shapes and no overlaps*. The women *argue* about *what shapes they can use*

Translation

Translation and Reflection
(Glide or footprint)

Two reflections
or reflection and translation

Two half turns
or half-turn and translation

One reflection and one half turn

One reflection and one translation
or reflection and glide

Three reflections

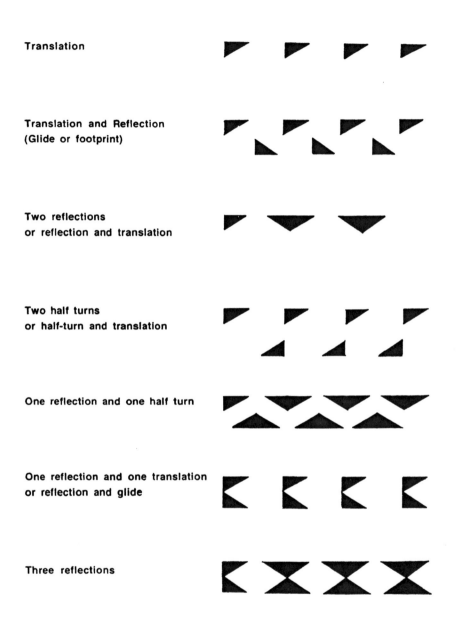

Figure 14.5. Transformations combining reflection and translation

Cross Stitch

Smile **2145**

You will need 5mm square paper.

Simple motifs, using cross stitch, can be used to create elaborate patterns by using transformations including reflections, rotations and translations.

The patterns below have been created by transforming this motif.

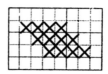

■ Analyse and describe each pattern in terms of transformations used.

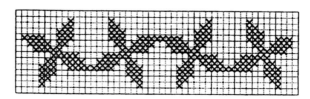

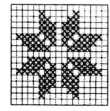

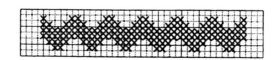

Turn over.

Here is another motif.

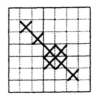

⊠ Analyse and describe the patterns created.

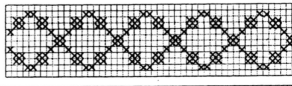

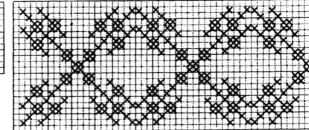

⊠ Using 5mm square paper to represent the cloth, design a motif of your own and create patterns, describing them in terms of reflections, rotations and translations.

⊠ Take one of your patterns and make it into a rectangular border keeping the design continuous and making right-angled bends in the appropriate places. You may find a mirror helpful to plan the corners.

You might like to choose one of your patterns and using cross stitch embroider it.

Adapted from an activity in M/W Cabbage.
© RBKC SMILE 1994.

Activity 2: "Cross Stitch." © SMILE Mathematics (con't.)

because *not every shape* will fit together like this. Will *shapes of three sides* work? Will they only fit if *all the sides are the same length*? What about *shapes of four sides? All shapes of four sides? Five sides? All shapes of five sides? No shapes of five sides?* And so on. (Clearly, some shapes allow more interesting or exciting designs than others. Thus, the group may need to search through a range of possibilities to find shapes that enable the construction of a unique yet practical design.)

Eventually they choose one *shape of four sides* and decide to make the quilt using *this shape only*. A mathematician would have called it a *concave quadrilateral* but the women call it an *arrowhead* because it is like a motif that appears in their church building. The women know that when they come to sew the *shapes* together, *small angles* are very fiddly and can result in the finished cloth being weaker in some places than others if they are not sewn together very skillfully. For this reason they have chosen an *arrowhead shape whose angles are all large enough* to sew comfortably. Before they begin to make the quilt the women first make *templates* from thin wood, plastic, or metal. These have to be very *accurate* because they are going to be *drawn around repeatedly* to produce *identical shapes* for sewing. There is bound to be some error in *drawing around a shape*, and a template that is inaccurate to begin with would *increase the amount of inaccuracy* in the final piece of patchwork.

Mathematical Analysis of Story 3

In mathematical terms, the women have been exploring *tessellation*, the properties of polygons that allow them to cover a plain surface completely without overlapping and gaps between them. The angle surrounding a point is 360°. Shapes that can fit together so that their corners meet with a total of exactly 360° will do so without overlapping and gaps. In other words, shapes will tessellate if their points meet at 360°. For example, squares tessellate because the right-angled corners of four of them fitted together total 360°.

In their explorations the women have been investigating properties of polygons that permit them to tessellate. This involves:

- Investigating regular and irregular triangles
- Investigating regular and irregular quadrilaterals
- Investigating regular and irregular pentagons, hexagons, and so on
- Drawing, measuring and calculating angles and sides of regular and irregular shapes
- Measuring and making very accurate angles

Suggested Classroom Activity Based on Story 3

Instructions. Design a patchwork quilt. One will need pencil, paper, ruler, compass and protractor, and a collection of interesting shapes that are traced, cut, or photocopied from magazines, books, posters, and the like.

- Explore different shapes suitable for making patchwork using the shapes that have been collected or made. Be systematic in the exploration
- Draw or make a collection of different triangles, regular and irregular. Draw examples of each and find ways of fitting them together to see if they tessellate. Make notes of the investigations
- Repeat the exercise with quadrilaterals, pentagons, and so on, always keeping records of what is being done
- Summarize the discoveries and deduce a rule for making shapes suitable for tessellated patchwork
- Try to make tessellations of two different shapes, being systematic in the search and deriving a rule for those that work
- Look for and describe geometrical patterns in floor and wall tiles, brick and stone walls, carpets and rugs, and so on

One can read about Marjorie Rice, a Californian housewife who became interested in mathematics while helping her children with their homework (see Schattschneider, 1981). She began a systematic study of pentagons (five-sided figures, some of which will tessellate and some not) and as a result reopened an area of higher mathematics that had been thought to have been solved. A good way to know more about tessellation is to begin with the work of the Dutch artist M. C. Escher (e.g., Ranucci & Teeters, 1977) or by looking at books on Islamic art and architecture.

DISCUSSION

I now return to the social and political issues surrounding the three questions presented earlier: the reasons why the obvious mathematical activity in women's work is not featured in mainstream mathematics education, and how it can be featured.

Why Isn't All This Mathematical Activity Featured in Mainstream Mathematics Education?

Effect of class on education. There has always been a tension between mathematics education for utility and mathematics education for its own sake. In the West, and dating back to at least the Greeks, there has been a social distinction between the mathematics done in institutions of learning and that done in places of physical work. This was a cultural decision, a matter of choice, and was unrelated to the cognitive demands of the mathematics (Harris, 1991b). For example, the same social distinction does not arise in Indian culture (Joseph, 1991).

The first attempts made in England and Wales to establish national criteria for school education were in the middle of the 19th century. They arose from the political need to educate the masses rather than from humanistic motives. If even a section of the workers were to be given the vote, then they had at least to be literate. These politically necessary educational moves were made at a time when distinctions by social class were widely believed to be natural, when there was genuine fear of working class revolution, and when the position of women of all classes of society was seen as domestic. Elementary education, which meant education for the working class, applied to six-sevenths of the population by government estimates and was carefully limited to basic literacy and preparation for the work children would enter on leaving school. The different curricula for different social strata of children concentrated on the three Rs but were most clearly differentiated in arithmetic and mathematics.

For the working-class children in the elementary schools, a minimum of arithmetic was deemed necessary for the work to which the children were to aspire, and no more. Even this minimum was only established after a period of heated debate in which strong lobbies argued against teaching the workers' children any arithmetic at all for fear that it might also give them the ability to question their station in life (Williams, 1961).

The curriculum for the middle class was also stratified. A summary of aspects of the Taunton Report of 1868 (Cooper, 1985) illustrates this clearly:

> The Report carefully classified education into three "grades," each appropriate for the children of a particular social grouping. The first, continuing to eighteen or nineteen years of age, was for the sons of "men with considerable incomes independent of their own exertions, or professional men, and men in business, whose profits put them on the same level" plus "the great majority of professional men, especial-

ly the clergy, medical men and lawyers; the poorer gentry; all in fact, who, having received a cultivated education themselves, are very anxious that their sons should not fall below them." These two subgroups were both seen as wanting more mathematics in their sons' mainly classical curriculum. The second grade, continuing to about sixteen, was seen as fit for the children of the "mercantile classes." Generally, for these, a curriculum less classical in orientation and with more stress on the practical was both seen as wanted and was recommended. Here, rather than mathematics, we find a reference to "arithmetic [and] the rudiments of mathematics beyond arithmetic." The third grade, appropriate for those staying on at school until about fourteen and being the sons of "the smaller tenant farmers, the small tradesmen [and] the superior artisans" should include "very good arithmetic" as part of a curriculum based on the three R's. (p. 36)

For the very small, élite minority of the "public" schools, there was Euclidian, pure mathematics learned for the purpose of exercising the brain, certainly not for practical use (Howson, 1982). The social grading of mathematics education by content was clear and intentional, with social status firmly linked to that content. And there was a clear social distinction between arithmetic and mathematics.

   Gendering mathematics education. In the society that graded children by mathematics education, and as seen clearly in the passage just quoted, "child" meant "boy." (Until very recently in mathematics education it still did—see later.) The Victorian consciousness of that age was one of transition in which there were widespread doubts about the nature of man and society; security and continuity thus lay in confidence in bourgeois industrial society and the place of the social classes and sexes within it (Houghton, 1957). The existing social order was necessary for a stable society and was reaffirmed throughout the century. Although the role of women was becoming contended, the submissive, domestic angle was the most widely accepted. There was a reaffirmation that women's delicate brains could not and their delicate bodies should not be subjected to the rigors of mathematics and science (Harding, 1986). Woman's role in all societal classes was domestic and girls' education had to prepare them for that role.

   A continuing characteristic of the education of girls was needlework, although the justification for teaching it, in fact, varied throughout the period and well into this century (Harris, 1997). Although there are records of boys being taught to knit, sewing was for girls only. It was frequently taught in the afternoon periods when boys did other subjects, including more arithmetic. Because the girls were taught less arithmetic than boys, less was expected of them in terms of both quantity and quality.

Mathematics and arithmetic were not only gendered but socially graded. Education reports of the time stress the nature and detail of school needlework, which for the masses had to be plain and undecorated. This was in contrast to the needlework of middle and upper class girls, who were expected to indicate their status and wealth by doing needlework that was purely ornamental (Parker, 1984). Interestingly, it was at this time that the inefficient but elegant style of knitting (sometimes called the English style) that displays pretty fingers to great advantage, was invented. The much quicker "continental" or "peasant" method used by people who actually needed to knit swiftly involved movement of the whole arm—far too coarse for the drawing room (Rutt, 1987). In the period in which masculinity and social class were differentiated through arithmetic and mathematics, femininity and social class were differentiated through both mathematics and needlework (Harris, 1997). A more recent review is by Leder (1992).

A rich body of literature and research on gendering in mathematics (summarized excellently by Willis, 1989) shows that although attitudes are changing, there is still in public consciousness, and across many cultures, a firm idea that mathematics is a male subject and that women and girls cannot do it. Until very recently, examples in school textbooks were almost invariably addressed to boys. For example, a survey of books covering the age range 3-13 published in England between 1970 and 1978 was conducted by Northam (1986). Of one of the books Northam reviewed she wrote:

> The visual material contains a number of adult figures, cowboys, soldiers, sportsmen, clowns, pedestrians and tractor drivers; they are tall, short, fat, thin, they climb ladders, they mend roofs, and they are almost all men. Two examples only depict women. Mr. and Mrs. Smith are seen walking to church and two female heads are seen in relation to three hair ribbons. (p. 114)

In her summary Northam noted:

> There is a clear tendency in the books studied to define mathematics as the province of males, and especially adult males. . . . Adult women are largely absent from these books, and by the age of thirteen girls have joined them in near-oblivion. There is an interesting parallel between the decline in girls' involvement in maths between seven and sixteen years of age, and the gradual disappearance of girls from maths books over the same period. (p. 116)

The examples quoted by Northam are of some of the social pressures that ease girls out of mathematical involvement from their earliest school days. A more detailed analysis of forces on girls over extended time at school is given by Walkerdine (1989).

The historical period that formalized the gendering of the subjects of mathematics and needlework was also the culmination of the period of increasing definition and differentiation of men's and women's work in factories, notably in the textiles industry (Holland, 1991). The designation of work done by women as unskilled because it is done by women (and for no other reason) still goes on in the industry, as do the low pay and conditions that follow from the designation (Cockburn, 1985).

Consequences. The legacies of gendered mathematics and gendered work, paid and unpaid, remain deeply influential in both explicit and hidden education curricula and in the social pressures on students with particular destinations. The result of these cultural pressures on girls and women is a widespread feeling among perfectly able females that mathematics is not for them and that it is irrelevant to their daily work or life.

Mathematics itself is powerful in two senses. Its axioms, symbols, and theorems are *internally* powerful in that within the subject of mathematics they represent an abstract system of precise meanings, explanations, and accepted truths that can be applied to other circumstances both inside and outside mathematics. This is what mathematicians mean when they talk about the power of mathematical modeling. But mathematics is also *externally* powerful. Since before the beginning of IQ tests, an ability to do various levels of mathematics has been taken as a differential mark of intelligence. Since IQ tests, the relationship has been formalized and quantified. The use of hierarchical layers of mathematics as an entry qualification for jobs, that is, the use of mathematics as a gateway, gives the subject power in controlling entry into the professions and economically powerful jobs in industry and commerce.

People who believe they cannot do mathematics are doubly powerless. They lack the basic steps and confidence to climb up the hierarchies in work and the entry key to power jobs. In my own research I have visited clothing stores and interviewed management and workers. In one branch of a well-known chain store, I asked about the promotion prospects of the women sales staff and was told, "The garment industry runs on percentages. These women will never understand percentages. That's why they will never be promoted." The textiles industries of the world depend on the cheap labor of uneducated women whose delicate fingers are still often quoted as a criterion for employing them (Elson, 1983). The same criterion may be quoted in recruiting low paid women into other industries, notably electronics, hence parallel gendered power structures are clear across many industries. The repetitive jobs that support the basic processes tend to be done by women. Those jobs in which the decisions about what processes are to be done and the systems that control the machines that do them are made, tend to be held by men (Cockburn, 1985).

The answer to my first question then, is that the mathematical activity in women's traditional work does not feature in mainstream mathematics simply because of the social expectations of both mathematics and of women's work in two particularly highly gendered fields. The work is not perceived as containing mathematics because nobody expects it to.

How Can the Mathematics of Women's Work Be Featured?

In 1987 I set up a small exhibition, Common Threads, in which every exhibit was a piece of sewn, knitted, embroidered, woven, or plaited textile of both domestic and industrial manufacture. All the captions, however, were mathematical. Each textile item was described mathematically to illustrate mathematical thinking that had gone or could have gone into making it, or the mathematics that could be illustrated by it. My aim in designing the exhibition in this way was to interweave the low-status activities of needlework, knitting, weaving, and so on, with the high-status language of mathematics so that each needed the other to make sense. There was a joint context for both activities that was dependent on an interaction between equals. By planning it like this I confronted stereotypes in both mathematics education and women's work. The levels of mathematics presented in the exhibition ranged from infant school to undergraduate, and the range and depth of mathematics displayed was enough to suggest that it would be perfectly possible to teach an entire secondary school course of mathematics through the medium of some kind of needlework. The judgment as to whether this would be advisable or appropriate was left to viewers. But the implication was that if a school mathematics curriculum based on feminine activities alone would be unbalanced, what could be said for one based totally on male activities?

The success of the exhibition in its subsequent tours of England and later the world under the auspices of the British Council, lay in the way Common Threads spoke immediately to both mathematics and women's educators. There is a familiarity with cloth, a welcome warmth to it that speaks to everyone without fear or favor. Mathematics presented through cloth rather than on paper carried none of the negative images of mathematics with it. In parallel with the exhibition I also published the pack of learning materials from which Activities 1 and 2 cited earlier were taken. This pack, called *Cabbage*,[1] also presented a challenge to traditional teaching methods in mathematics education by the methods it assumed.

A strong tradition in mathematics teaching, particularly in the lower levels of vocational education and adult education is that of teaching particular skills in class, often through rules, and then applying them to examples in the work or other context for which the training is aimed. Such teaching concentrates on what has come to be called *instrumental understanding*, the rote learning and use of rules and procedures, at the expense of *relational understanding*, the understanding of principles and meaning (Skemp, 1976; see chap. 2, this volume). A major problem of the instrumental "skills and applications" approach is that it is generally ineffective; the skills remain specific to the situation in which they were learned. The phenomenon of students who can do mathematics in class but cannot do the same mathematics outside it is well-know in education.

Strässer, Barr, Evans, and Wolf (1991), in discussing problems of teaching for skills versus for understanding, review research in a training scheme (program) in which conditions were varied for the students in terms of the range and number of contexts to which students' learning was applied. The results consistently showed that the use of varied contexts encouraged the development of skills that did generalize. Variation of contexts inside the classroom and in the workplace did encourage the development of general understanding in a way that concentrating on repeated routine applications did not. The trainees who made the least progress were those who had no prior experience of the problems they were asked to solve. Strässer et al. (1991) concluded that "variation of contexts . . . tends to encourage the development of general understanding in a way which concentrating on repeated routine applications of algorithms does not and cannot" (p. 163).

---

[1]The word *Cabbage* comes from old French *coupage* and refers to a practice that is rather frowned on in the clothing industry. When a manufacturer gives out a precise amount of cloth to be made into a specific number of garments, the workers can sometimes make more garments, which the manufacturer does not know about, by careful placing of the garment pieces. The practice becomes disreputable when the workers cut down on seams or place pieces at an incorrect angle (which reduce the quality of the original garments) so that they can make more cabbage. I called the pack *Cabbage* in honor of the unsung skills of garment workers. The Sweater and Cross Stitch activities were designed by the author and appeared in *Cabbage*, an instructional package now out of print. However, these activities were also incorporated into the SMILE Mathematics program (or scheme, in U.K. terms). SMILE can be obtained from the SMILE Centre, Isaac Newton Centre for Professional Development, 108A Lancaster Road, London W11 1QS, England. Figures reproduced with permission of the publisher.

Behind this instrumental, skills approach to learning is the assumption of a process of *transfer*: the application by the learner of concepts and skills learned in one place to the conditions of another. The research described earlier, taking the ability to generalize as a measure of transfer, suggests that the idea of transfer is a complex one, affected by the range of contexts within which particular learning is taught. It is but one example of research that is critical to our understanding of the idea of transfer.

Other studies such as those of Lave (1988) argue that an assumption of transfer is a necessary feature of behavioral models of teaching and learning. To quote Lave, "Conventional academic and folk theory assumes that arithmetic is learned in school in the normative fashion in which it is taught and is then literally carried away from school to be applied at will in any situation that calls for calculation" (p. 4). But transfer is an assumption, not a finding. Lave's view is that the mathematics of formal education and that of workplaces is presently better characterized not by continuity through an idea such as transfer, but by discontinuity. The mathematics of formal education and of workplaces is different in the way it is understood and practiced. A philosophical justification for this view is provided by Dowling (1991). Work summarized by Carraher (1991), summarizing previous research, gives examples of some of these different practices and meanings. The relatively new field of ethnomathematics provides many more examples. Bishop (1991) summarizes much of this work and concludes that all cultures derive their own mathematics to suit their needs. A more recent review is by Gerdes (1997).

The major virtue of self-invented or specific mathematics is its meaningfulness. Its major liability is the limited conditions to which it can be applied. The educational problem then becomes that of developing those meanings without discrediting or destroying the self-confidence of the learners, while at the same time helping the learners to come to understand the symbols, conventions, and meanings of the internationally recognized mathematics of formal education. Such a view of learning suggests *transition* rather than transfer. The educational problem is neatly summarized by Carraher (1991) as "making meaningful representations and procedures more powerful on the one hand, and making powerful representations and procedures psychologically meaningful [on the other]" (p. 195).

The learning materials incorporated into the teaching pack *Cabbage* were aimed at addressing these issues in teaching and offering contexts more likely be familiar to female students. The activities were mainly practical and involved the exploration of contexts in which mathematics was one, albeit an important feature. *Cabbage* had been designed in the same format as a previous pack that had used cardboard as the context of practical design activities that included mathematics. In a discussion of its uses I wrote:

> There is no lack of arithmetic and the better you can do it the better you can get on with the problem. The point is that the arithmetic is just one problem-solver's tool; so is measurement, argument and the skill of handling things in three dimensions. But just as important as the tools themselves is the way people learn to use them. (Harris, 1985, p. 46)

The effectiveness of this approach is exemplified by some classroom work by Lolley and Ross (1989). They set up a mathematics project in a girls' secondary school, on making a patchwork quilt, originally with the aim of covering four syllabus items: the rectangle, angles, the triangle and rhombus, and symmetry. Lolley, although an experienced and successful mathematics teacher, was nervous about shifting to activity-based teaching for fear of not covering the syllabus in time for the examination. However, she felt the need to change her teaching methods because, although her girls were successful in examinations, she believed that they were not retaining what they had learned, that they had not made mathematics their own. She had also noticed, in confirmation of research findings, that girls often worked better in a group than singly. As the project developed and as the girls worked with each other on a theme of patchwork, their own enhanced motivation led them to explore much more mathematics than the four topics Lolley had planned to cover. In addition to the intended syllabus items the class explored which polygons will tessellate, how many triangles there are in a polygon, what happens when you rip off the corners of polygons and stick them together, drawing polygons by dividing circles, measuring interior angles of polygons and adding them up, and the symmetry of polygons. The pupils directed their own learning and sought new mathematical terminology when they wanted it; they were able to explain their mathematics in everyday language and their learning was generalized.

There is no doubt that the mathematics in traditional women's work can feature in mathematics education through relevant activities and teaching methods that are known to favor girls. There is also no doubt that the introduction of textile activities into traditional curricula can enrich both content and teaching for all students by redressing current imbalances in both content and teaching.

Why Should We Feature the Mathematics of Women's Work?

The problems discussed earlier arise within the context of mathematics perceived as a given, an abstract body of knowledge somewhere out there that has to be learned and applied. "For over two thousand years, mathematics has been dominated by an absolutist paradigm, which views it as a body of infallible and objective truth, far removed from the affairs and values of humanity" (Ernest, 1991, p. xi; and see Coben's account of this issue in chap. 2, this volume). Given this, it is ironic that a subject that is supposed to be so abstract and remote from human frailty should be the one above all others that generates such human feelings as fear or even panic about the learning of it (Buxton, 1981). The effects of the absolutist view in practice is that the values within it serve the interests of a privileged group that advantage males over females, whites over blacks, and middle classes over lower classes, in terms of academic success and achievement (see Ernest, 1991; this argument is too detailed to summarize here).

But mathematics itself has been undergoing a philosophical revolution. It is increasingly being recognized as a product of human inventiveness which, like other human inventions, takes place in particular times in particular social settings. If mathematics is seen as a body of infallible, objective knowledge then it follows that:

> the under-participation of sectors of the population, such as women; the sense of cultural alienation from mathematics felt by many groups of students; the relationship of mathematics to human affairs such as the transmission of social and political values; its role in the distribution of wealth and power; none of these issues are relevant. (Ernest, 1991, p. xii)

If, however, mathematics is seen as a social construction, it is revealed as an ongoing process of inquiry, an expanding field of human invention, an unfinished product subject to human fallibility.

> Such a dynamic view of mathematics has powerful education consequences. The aims of teaching mathematics need to include the empowerment of learners to create their own mathematical knowledge; mathematics can be shaped . . . to give all groups more access to its concepts, and to the wealth and power its knowledge brings; the social contexts of the uses and practices of mathematics can no longer be legitimately pushed aside, the implicit values of mathematics need to be squarely faced. When mathematics is seen in this way it needs to be studied in living contexts which are meaningful

and relevant to the learners, including their languages, cultures and every day lives. . . . This view of mathematics provides a rationale, as well as a foundation for multicultural and girl-friendly approaches to mathematics. Overall, mathematics becomes responsible for its uses and consequences, in education and society. (Ernest, 1991, p. xii)

This approach to mathematics education has been taken up and developed by workers concerned with what they are calling "criticalmathematics," the definition of which is currently being closely argued. Members of the criticalmathematics education group (notice the use of a single word) such as Frankenstein (1989) place the teaching and learning of mathematics within a rationale that links education to considerations of citizenship and social responsibility. Frankenstein's teaching of mathematics actively uncovers the political and social uses to which it is put in a controlling society. Her students become active critics of, and participants in, rather than victims of society, learning both mathematics and the abuses to which it is put at the same time.

Caveat. Throughout this chapter I have been deliberately and overtly working with stereotypes, presenting mathematics as the stereotype of masculinity and needlework as the stereotype of femininity. Both are stereotypes: they do not represent more than that. There is no male mathematics or female needlework mathematics. There is mathematics, a body of rich, powerful, and fascinating knowledge and processes that is everyone's birthright. There is also textiles work, gendered in some cultures but not in others. In Senegambia, for example, weaving is a male occupation, credited with intellectual and power values that are regarded as beyond the capabilities of women. In a community in Nigeria, where very similar weaving is done by both men and women, the activity does not carry such powerful value loadings (Harris, 1987).

My aim in concentrating so much on the mathematics in stereotypical women's work is to expose both stereotypes for what they are. My teaching activities, which start from female stereotypes, are not aimed at glorifying women's self-produced knowledge, for this can be a deeply patronizing, disempowering, and ultimately self-defeating way of proceeding. They can, however, show how deeply gender values can be held and how damaging they can be. In using such activities with mixed groups of teachers in training I have been astonished by the depth of feeling shown by some males who regard these activities as degrading (Harris, 1991c).

The justifications for featuring traditional women's work in mathematics education are social and political. The dual aim is for mathematics education that reflects mathematical activity in society in general more accurately than it does at present and for a mathematics education that recognizes women as people with their own valid intellects.

IMPLICATIONS AND SUMMARY

The justification for starting mathematical activity from traditional women's work when appropriate or desirable has been given earlier. The following three examples illustrate how the implications of the line of thinking developed here have been taken up in three different educational contexts, involving issues of cultural confidence, economic development, and educational enrichment: First, those in ethnic minorities who may have experienced formal mathematics education in unfamiliar contexts arising from ideologies and cultures foreign to them or who, because of their minority status, have been expected only to have low achievements. Second, within the field of development education, I discuss the case of women being taught textile crafts for reasons of economic independence, who could simultaneously be accredited for the mathematics embedded in what they are already learning. Third, I describe current work on a project in England aimed at raising mathematics and science awareness and understanding within the community of a national women's organization.

Cultural Confidence

While the teaching pack *Cabbage* was being developed, my attention was drawn to the particular needs of groups of children from parts of Bangladesh who had recently immigrated to parts of London where there was already a Bangladeshi community. Such immigration patterns can occur quite suddenly in schools through the arrival of groups of children often traumatized by culture shock, and who speak very little English. Without particular educational efforts by the host schools, such children can remain isolated. The resulting loss is two-way, in what the children need from their host community so that they can survive and thrive and in what they can contribute to their host community from their own culture and skills. In this two-way interaction lies the rich possibility of an integration of learning starting from what people are already doing.

At about the time of arrival of a particularly large group of children, there was an exhibition of Bangladeshi crafts in London to which I went, knowing that it would be full of mathematical activity, particularly within weaving. I selected some items including circular fans which are woven from flat bamboo strips in such a way as to reveal different patterns. My first draft materials involved repeating the weaving patterns with strips of paper. These were rejected by the teachers who were, unusually, all English, on the grounds that they could not weave and therefore could not do the activity in class. I returned the rejected drafts with the suggestion that they should ask the children to show them how to weave. The result was an immediate and very positive, two-way interaction. Doing something as familiar as weaving "from home," the children demonstrated a respected skill that neither staff nor other pupils had; in teaching their peers how to weave their language skills and confidence grew in parallel with the skills of their peer group; and, under the guidance of the teachers (now trainee weavers themselves), both groups worked from different cultural bases on the geometry of tessellation that formed the weaving patterns.

Economic Development

In many developing countries, self-help and funded vocational programs aim at teaching textile crafts to girls and women so that they acquire marketable skills. While the *Common Threads* exhibition was touring I visited many such groups in several African countries. In one country I spent a half day in participatory observation of a needlework class for young women and another half day in a woodwork class for young men in a vocational training institution whose students were recent school graduates. In the woodwork class the arithmetic needed for the tasks was explicit and openly discussed between students. There was no problem with the fact that the craft needed arithmetic and that arithmetic had to be done within it. The social setting was entirely male, and the work was confident and good.

In the needlework class, students were working from worksheets taken from a text book that instructed them how to make particular necklines. When there was a problem, for example, when some students did not understood the instructions about taking a measurement from a midpoint, the tutor taught by demonstration and rules, such as, "You have to put the end of your tape just here like this." Both students and tutor expressed lack of confidence in doing arithmetic, although during the actual practice of the needlework they frequently did it without problems. The arithmetic expected of them in their task was of wider variety and deeper level than that of the woodwork class. After analyzing the arithmetic in both classes and exploring of the craft curricula for both courses, it was clear that in the needlework course there was enough range and depth of arithmetic implicit in the work to identify and accredit it as a separate numeracy certificate. There would, of course, be teacher training implications, but they could be taken up later.

In this particular instance, funding is now being sought to develop such a course and accreditation. The aim is to build on what the women are already doing for the purpose of awarding two qualifications, one a craft certificate that will enable them to get a job or be economically active in the domestic sphere of textiles, the other a numeracy certificate or qualification that would enable them to "buy in" or be able to apply for jobs of higher status such as accounting, or go back into the education system at a higher level. At the time of writing no response had been received to the aid application but there was fear that the idea would be too radical for traditional-minded funders. Such formal bodies still tend to hold traditional views of mathematics education and traditional views of women's work.

Educational Enrichment

A recently launched program in England and Wales aims to make more popular the understanding of science and mathematics. There has been a particularly positive response to the initiative by the Women's Institute, which currently has about 250,000 members. The Institute has its own college, where a very wide range of short courses are taught. Under this scheme (program) I was invited to explore ways in which mathematics could be developed in college courses. It was agreed that I would attend three textile courses as participant-observer and report on any mathematics I found that could be developed in courses. The three courses I attended were on curtain and blind (shade) making, knitting, and machine-made patchwork.

My most salient finding was that in all three courses, every item of on-task talk contained mathematics. Every activity in which instruction was asked, given, or shared, involved measurement, calculation, or symmetric relations of some sort. The next most obvious feature was the widespread negative attitude to mathematics once its presence had been acknowledged by the students. This was extremely difficult to handle because the attitude was often combined with feelings of real anxiety or distress. In conjunction with the tutors I found that discussing mathematical things as they occurred (and I was an active student of the course, sewing and knitting alongside the other students), including my own lack of understanding of the ways in which some of the craft calculations were done, did help to relieve feelings of inadequacy in the group.

At the time of writing this chapter, a course on Christmas decoration had just finished. In this I worked with a scientist colleague who gave a 1-hour presentation on the physics and chemistry of snowflakes as an explanation of why they have six points. This was followed by my hour, in which I developed the theme of six into different ways of making stars from hexagons. While awaiting formal feedback from participants, the high level of positive verbal feedback was noted. In this course, the concession to gender was in developing the considerable practical experience of the women in running households (most of the women were over 50 years of age) through the theme of decorating the Christmas dinner table with hexagons. In both the science and mathematics part of the course, the teaching style was open-ended and interactive, with numerous suggestions for individual development.

Summary. In all three cases discussed in this section, mathematics learning took place starting from and within contexts that were meaningful and pleasurable to the learners. This has always been a characteristic of good teaching of anyone. The excitement involved in starting with textiles comes out of their ubiquity. We all wear clothes and those clothes have shape, form, symmetry, and decoration. Getting into the habit of looking through mathematical eyes at what we put on in the morning (check out the construction of ties, socks, or shoes), or use to decorate our house, can be a richly rewarding experience for both teachers and learners.

More broadly, we should recognize the extent to which mathematics and mathematics education have been and still are gendered and realize how damaging this can be to very large numbers of potential learners. Heightened awareness of processes of trivialization and devaluation of meaningful mathematics embedded in traditional crafts and in women's activities can help plan a very positive beginning to a fresh start for learners and teachers, male and female alike. Teaching mathematics that builds from what people already do, such as needlework, can validate the experience of women, reverse the undeserved feeling of many women that they are not able to do mathematics, be a source of legitimation of numeracy, and enrich the whole context of mathematics education.

REFERENCES

Bishop, A. (1991). Mathematics education in its cultural context. In M. Harris (Ed.), *Schools, mathematics and work* (pp. 29-42). New York: Taylor & Francis.

Buxton, L. (1981). *Do you panic about maths?: Coping with maths anxiety.* London: Heinemann Educational.

Carraher, D. (1991). Mathematics in and out of school: A selective review of studies from Brazil. In M. Harris (Ed.), *Schools, mathematics and work* (pp. 169-201). Basingstoke and New York: Falmer Press and Taylor & Francis.

Cockburn, C. (1985). *Machinery of dominance: Women, men and technical knowhow.* Dover, NH: Pluto Press.

Cockcroft, W. H. (1982). *Mathematics counts: Report of the Committee of Inquiry into the Teaching of Mathematics in Schools under the chairmanship of Dr. W H Cockcroft.* London: Her Majesty's Stationery Office.

Cooper, B. (1985). *Renegotiating secondary school mathematics: A study of curriculum change and stability.* New York: Taylor & Frances.

Dowling, P. (1991). The contextualising of mathematics: Towards a theoretical map. In M. Harris (Ed.), *Schools, mathematics and work* (pp. 93-120). New York: Taylor & Francis.

Elson, J. (1983). Nimble fingers and other fables. In W. Chapkis & C. Enloe (Eds.), *Of common cloth: Women in the global textiles industry.* Washington DC: Transnational Institute.

Ernest, P. (1991). *The philosophy of mathematics education.* New York: Taylor & Francis.

Frankenstein, M. (1989). *Relearning mathematics: A different "R"—radical mathematics.* London: Free Association Books.

Gerdes, P. (1997). Survey of current work in ethnomathematics. In A. B. Powell & M. Frankenstein (Eds.), *Ethnomathematics: Challenging eurocentrism in mathematics education* (pp. 331-371). Albany: State University of New York Press.

Harding, J. (Ed.). (1986). *Perspectives on gender and science.* New York: The Falmer Press.

Harris, M. (1985). Wrapping it up. *Mathematics Teaching, 113,* 44-146.

Harris, M. (1987). Mathematics and fabrics. *Mathematics Teaching, 120,* 43-45.

Harris, M. (1991a). Looking for the maths in work. In M. Harris (Ed.), *Schools mathematics and work* (pp. 132-144). New York: Taylor & Francis.

Harris, M. (1991b). Mathematics in context. In M. Harris (Ed.), *Schools, mathematics and work* (pp. 1-12). New York: Taylor & Francis.

Harris, M. (1991c). Postscript: The maths in work project. In M. Harris (Ed.), *Schools, mathematics and work* (pp. 284-291). New York: Taylor & Francis.

Harris, M. (1997). *Common threads: Women, mathematics and work.* London: Trentham Books.

Holland, J. (1991). The gendering of work. In M. Harris (Ed.), *Schools, mathematics and work* (pp. 230-252). New York: Taylor & Francis.

Houghton, W. E. (1957). *The Victorian frame of mind.* New Haven: Yale University Press.

Howson, G. (1982). *A history of mathematics education in England*. New York: Cambridge University Press.

Joseph, G. G. (1991). Foundations of Eurocentrism in mathematics. In M. Harris (Ed.), *Schools, mathematics and work* (pp. 42-56). New York: Taylor & Francis.

Lave, J. (1988). *Cognition in practice*. Cambridge: Cambridge University Press.

Lave, J., Murtaugh, M., & De la Rocha, O. (1984). The dialectic of arithmetic in grocery shopping. In B. Rogoff & J. Lave (Eds.), *Everyday cognition: Its development in social context* (pp. 67-94). Cambridge: Harvard University Press.

Leder, G. C. (1992). Mathematics and gender: Changing perspectives. In D. A. Grouws (Ed.), *Handbook of research on mathematics teaching and learning* (pp. 597-622). New York: Macmillan.

Lolley, M., & Ross, K. (1989). The patchwork quilt. In M. Harris (Ed.), *Textiles in mathematics teaching* (pp. 18-21). London: University of London Institute of Education, Maths in Work Project. (Out of print)

Miller, J. (1991). Textiles and symmetry. *Mathematics Teaching, 137*, 26-27.

Northam, J. (1986). Girls and boys in primary maths books. In L. Burton (Ed.), *Girls into maths can go* (pp. 110-116). Eastbourne: Holt, Rinehart & Winston.

Parker, R. (1984). *The subversive stitch: Embroidery and the making of the feminine*. London: The Women's Press.

Ranucci, E. R., & Teeters, J. L. (1977). *Creating Escher-type drawings*. Synovial, CA: Creative Publications.

Rutt, R. (1987). *A history of hand knitting*. London: Batsford.

Schattschneider, D. (1981). In praise of amateurs. In D. A. Klarner (Ed.), *The mathematical gardener* (pp. 597-622). Boston: Prindle, Weber & Schmidt.

Scribner, S. (1984). Studying working intelligence. In B. Rogoff & J. Lave (Eds.), *Everyday cognition: Its development in social context* (pp. 9-40). Cambridge: Harvard University Press.

Skemp, R. (1976). Relational and instrumental understanding. *Mathematics Teaching, 77*, 20-26.

Strässer, R., Barr, G., Evans, J., & Wolf, A. (1991). Skills versus understanding. In M. Harris (Ed.), *Schools, mathematics and work* (pp. 158-168). Bristol, PA: Falmer Press.

Walkerdine, V. (1988). *The mastery of reason*. New York: Routledge.

Walkerdine, V. (1989). *Counting girls out*. London: Virago Press.

Williams, R. (1961). *The long revolution*. Harmondsworth: Penguin Books.

Willis, S. (1989). *Real girls don't do maths: Gender and the construction of privilege*. Victoria, Australia: Deakin University Press.

IV

Assessment

15

Assessment in Adult Numeracy Education: Issues and Principles for Good Practice

Joy Cumming
Griffith University

Iddo Gal
University of Haifa

This chapter reviews changes in thinking about assessment in mathematics and numeracy education and outlines suggestions for changing practices. The chapter aims to provoke teachers to reflect on questions, such as Why are we assessing? What is good assessment? How best can we assess in adult numeracy education? and, pragmatically, How can we gain adequate assessment information in situations that are often very short, minimally funded, and externally driven?

INTRODUCTION

The increased public attention paid in recent years around the world to adult literacy and numeracy issues has been based in part on economic rationalism—tying higher skill levels to a more productive or competitive workforce and hence a better economy. It has also been supported by a growing acceptance of the position that education (and literacy) is a basic right of all individuals. This attention has been reflected in considerable increases in funding by governments and other organizations for literacy and numeracy programs, both general programs as well as those that aim to prepare learners for the labor market.

Although attention to literacy and numeracy has been growing, the *assessment* of the literacy or numeracy skills that people (students) have, can apply, or need to further develop, has become a major challenge for those involved in adult education. This challenge is caused by the confluence of several interrelated processes:

1. Theoretical developments over the last two decades (Venezky, Wagner, & Ciliberti, 1990) have increased our understanding of the multiplicity of contexts and pathways in which people's literacy and numeracy skills and dispositions can be developed and practiced.
2. New perspectives on the goals of mathematics education (e.g., NCTM, 1989) and workplace preparation in schools (SCANS, 1991) outline an expanded set of skills and dispositions as the targets of educational interventions that aim to prepare students for real-world functioning.
3. Advances in the theory and practice of educational assessment, as well as in literacy and mathematics education, demand more elaborate and authentic assessments that encompass the expanding range of skills, knowledge, and dispositions included in literacy and numeracy.
4. There is a growing emphasis on educational accountability, and demands from funders that adult education programs provide credible and informative data about changes in students' skills.

These processes present dilemmas for students, teachers, program administrators, and policymakers. Teachers and programs face a problem if they want to address a wide range of issues in their instruction, as well as expand their assessments within a limited time frame, and at the same time collect data about student performance that will satisfy the information needs of policymakers and funders.

This chapter examines recent changes in thinking about the assessment of mathematical skills and provides some suggestions for effective practices that can be incorporated into adult numeracy programs. (Chapter 16, this volume, illustrates an innovative approach to assessment of functional numeracy skills; Chapter 5 also touches on assessment issues.)

NUMERACY REVISITED

Let us first clarify what we mean by numeracy. In Chapter 1 (this volume), Gal argues that numeracy describes an aggregate of skills, knowledge, beliefs, habits of mind, and general communicative and problem-solving skills that individuals need to acquire to effectively handle real-world situations or interpretive tasks with embedded mathematical or quantifiable elements. The diversity of life contexts in which learners may need to apply numeracy skills and dispositions implies that it is essential to view numeracy as both relative and dynamic, rather than as encompassing a fixed set of skills. (In contrast, the traditional mathematics school curriculum attempts to impart to all learners the same computational skills and factual knowledge.)

Following Chapter 1, we argue that one can have knowledge of a mathematical fact, concept, procedure, or strategy, but it is in the appropriate selection and use of that mathematical knowledge *in purposeful contexts* that one demonstrates that one is numerate. With this in mind, we note that mathematical performance involves both domain-specific knowledge and strategies, as well as general cognitive skills (Perkins & Salomon, 1989; Sticht & McDonald, 1993) and related competencies that adults may have acquired outside the classroom. It is important to acknowledge the richness of *informal* mathematical knowledge, problem solving, and reasoning skills that adults may already possess and can bring to their studies. Bishop (1988), an ethnomathematician, identified six major categories of mathematical activities common across almost all cultures:

- counting
- locating
- measuring
- designing
- playing (games)
- explaining

In a study now underway by the first author (Cumming, forthcoming), of 160 participants, including school children from 5 to 15 years of age and adult numeracy students, only two participants reported that they did not play any form of game or that they could not explain rules of a game such as scoring or any strategies they used to win in a game.

Assessments should adopt a broad definition of numeracy to encompass the rich cognitive skills and knowledge used in everyday activities. Such assessments can show students that they "know" more mathematics than they thought they did, leading to an immediate instructional impact. Validation of students' informal knowledge can also assist in conveying to students (and to some teachers) that mathematics is not just the performance of school-based computations on which one is right or wrong or which one approaches in only one "right way."

The upshot is that in adult numeracy education there is a need to teach and, in turn, assess both formal mathematical knowledge as well as success in coping with tasks representing diverse functional contexts. This requires that a wide range of tasks and contexts be employed that can elicit the cognitive knowledge-base and reasoning processes, metacognitive strategies, dispositions, and selected literacy skills that jointly contribute to a person's numeracy. That said, the aspects of knowledge and performance to be assessed may depend on the purposes of the assessment, which is explored in the following section.

PURPOSES OF NUMERACY ASSESSMENT

Assessment is undertaken in adult education for many general purposes related to the learner, the curriculum, and the program itself:

Learner:	Initial diagnosis for placement
	Formative
	Summative
Curriculum:	Goal setting
	Planning
	Modification
Program:	Evaluation
	Reporting
	Accountability

Although these assessment-related terms are undoubtedly familiar to most teachers, in many adult education programs, assessment practices that teachers end up using are not differentiated in terms of these purposes because of a variety of factors. There is an overall tendency to use only one or two styles of information gathering for multiple purposes. Next we examine the implications of this situation for the major purposes for which assessments are conducted in numeracy education.

Initial Diagnosis to Inform Placement and Planning

Diagnostic assessment is undertaken to allow planning of instruction and placement. The value of diagnostic assessment for an adult learner is enhanced if it not only points to computations the learner can or cannot do, but identifies the mathematical experiences and knowledge that adult learners bring to a situation (such as those mentioned by Bishop and others) and learners' problem-solving and reasoning strategies. This provides a broader basis for designing educational experiences.

In Australia and the United Kingdom, where the adult literacy and numeracy sectors have emerged mostly from community-based, holistic philosophies of provision, many teachers are resistant to standardized forms of assessment. For initial "placement" purposes they may use a locally developed "sheet of 'sums'" (i.e., simple computational exercises for which the student has to generate answers), as well as qualitative observations. In the United States, the opposite trend appears to hold; over three-fourths of all U.S. programs rely on or are required by federal and local funding agencies to use a common standardized tool, such as the Tests of Adult Basic Education (TABE, see later); (Gal & Schuh, 1994), although some also use additional measures of a more qualitative or informal nature.

Both approaches in practice tend to emphasize computational activities similar to primary school arithmetic. However, the earlier discussion suggests that the use of a written computational test, even if not standardized, will uncover only some but not all of the many valuable skills and dispositions the student may possess. Although computational skills are not unimportant, students tend to overemphasize them. The fact that such skills are not the sole focus of a program, as well as the other aspects of numeracy that a program emphasizes, should be discussed with students. It is important to prevent an unwarranted perception that computations are the sole focus of mathematics learning for adults and place students' initial experience with any formal testing in a broader context.

To supplement written or qualitative assessments, incoming students could be asked to self-evaluate their reading and mathematics skills, point to problems or difficulties they may have in these areas, and describe their learning goals. It is important for the practitioner to maintain a broad concept of the nature of numeracy because the adult learner often will not. Just as many adult literacy students indicate that they want to learn to spell, many also express their learning needs in numeracy in terms of the math they did not learn at school (e.g., "I want to learn fractions"). A self-evaluation, although an unstructured and perhaps flawed screening procedure, may be needed because people who have decided they need, or are suspected to need numeracy (and literacy) instruction, may not perform well on any initial formal testing due to anxiety, low test-wiseness, being "rusted," or other factors. A self-evaluation screening procedure also has the advantage of not specifying in advance what numeracy content the student is to learn, allowing the opportunity to clarify a student's learning goals and expectations and to negotiate a curriculum.

Initial diagnosis in a workplace learning context requires special consideration and should be based on an examination of the actual vocational setting and on an audit of expected performance and needed skills. Such skills should be characterized based on task analysis (involving direct observation of performance, interviews with workers, and other steps), rather than only on a survey of employers regarding what (mathematical) skills they believe the job requires. These two types of data are often very different in reality and may lead to very different types of assessment (Taylor, Lewe, & Draper, 1987).

We should also examine *how* findings from an initial diagnosis are being used to inform placement and instruction, and what is the degree of "precision" required of the initial assessment. Recent work by Venezky and his colleagues at the National Center on Adult Literacy in the United States (Venezky, 1992) imply that programs often overassess during initial diagnosis when relying on standardized tests such as the TABE. Administration of the full TABE battery requires over 3 hours; however, students' scores on the TABE Locator test (a brief pretest with items representative of the full version that takes less than 30 minutes) were shown to be as predictive of future performance in the program as the full battery. Further, the Locator test was shown to provide enough information on the basis of which to classify students into one of 3-4 levels of instruction, which is the common type of initial placement decision. Time saved by using shorter standardized tests can be used instead to assess other key thinking processes and informal or practical mathematical knowledge of learners via alternative assessment methods.

On the issue of the implications of the information gleaned from initial diagnosis, researchers such as Willis (1990) have commented that school children may be able to perform certain tasks better than adults (e.g., construct a graph, perform a computation, memorize number facts), yet have limited understanding of the meaning of what they can generate. In contrast, more adults would have the ability to interpret a graph or number fact in relation to a real-world context or question, or understand its implications, even when having difficulty constructing it. A diagnostic assessment that establishes that an adult cannot generate a number or display may cause a teacher to identify this skill as a learning goal, but this may be misleading. Assessments need to also tap into adults' interpretive skills, not just their generative skills.

In summary, diagnostic assessment in adult numeracy education should encompass a broad definition of numeracy and seek to:

- Establish the learner's goals
- Determine what knowledge, strategies, and reasoning processes the adult learner already possesses, whether formally or informally (see Chapter 16, this volume)
- Determine what needs to be learned in the context of the learner's goals and the learning setting (e.g., a workplace)
- Be able to indicate which strategies should be used in instruction

Formative Assessment

Formative assessment should be undertaken on an ongoing basis to provide feedback both to the teacher and to the student on how the student is progressing, to identify areas that need redressing, and to inform decisions on where instruction might proceed. This type of assessment is used informally by many adult numeracy teachers, although many teachers are not aware that they are doing this. Formative assessment can involve a mix of formal and informal methods, including but not limited to the teacher talking with or looking over the shoulder of students, the use of extended group projects, brief written tests, simulations, and so forth.

The effectiveness of formative assessments is critically dependent on the definition of numeracy held by the teacher. Too many adult numeracy programs are still focusing on assessing students' basic computational activities, rather than also identifying how learned skills enable students to achieve life or work goals that they could not previously do, and on whether students' reasoning processes and their conceptual understanding (of the mathematical ideas behind any computational procedures they learn or use) are improving. Students depend on their teachers to teach and assess what is most significant to learn. Students working on narrowly focused math activities will assume that they are progressing on the most important aspects of mathematics. This places a professional responsibility on teachers when they design formative assessments, as they have to take into account and potentially resolve conflicts between students' declared goals (which may reflect their memories of what "school math" encompassed) and broader perspectives on what "being numerate" includes.

Appropriateness of the question and of the response. A subsidiary issue is that when the definition of numeracy is broadened, the definition of acceptable questions and "correct" answers is broadened as well. Questions and correctness are context dependent. Examples provided by numeracy authors such as Willis relate to standard school-type questions that purport to simulate real-life problem solving, such as "which is the cheaper coffee to buy?" In such a question, the larger container is meant to be the cheaper unit to purchase; however, in real life, if one does not drink coffee frequently, the coffee would go stale and would therefore not be a good purchase. Similarly, a question such as, "You want to buy two tickets for the movies; tickets cost $4.50 each; how much money should you take with you?", could elicit the response, "$20 because I like to have some extra money on me (e.g., for other movie-related expenses)." In real life, this answer demonstrates an effective (mathematical) strategy, although it would be considered wrong in a classroom context if the question poser intended to examine a students' addition skill without taking into account broader real-world considerations.

Conversely, if a person is undertaking numeracy training for a specific job, accuracy on some tasks will be essential. It is therefore important that judgments about accuracy and correctness are made with respect to contextual demands, rather than to formal concepts of accuracy (i.e., what is computationally correct). As in diagnostic assessment, in many formative contexts oral assessment and discussion of numeracy strategies and activities can be used. Through this it is possible to clarify the students' own understanding of the tasks being presented, become aware of any real-world constraints or beliefs that have influenced students' approach or performance, and only then decide if learners are gaining in mathematical knowledge and understanding and are able to apply their knowledge in new ways.

Summative Assessment and Reporting

Reporting to outside agencies on students' level of achievement at the end of a course of study is an area in which a lot of assessment occurs but in which theoretically, the most improvement is needed. What assessments are used for this purpose may depend on external program constraints. A certificated program requires evidence of progress toward common documented goals; individually focused programs, such as community programs and some work preparation programs, can also report individual progress toward idiosyncratic goals. The end result is that diverse forms of summative assessment now exist and are accepted as appropriate in different education sectors.

In general, instructors in all settings should undertake more systematic, documented, and appropriate assessment of their students' overall progress, after reexamining their definition of numeracy. In examining reporting systems in workplace education programs, for example, Mikulecky and d'Adamo-Weinstein (1991) found that few programs had structured assessments and that most relied instead on (problematic) anecdotal information. Effective programs, however, used highly structured assessments specifically designed for measuring change in skills identified as relevant in the particular workplace, rather than generic tests.

The need to provide data to external authorities to satisfy accountability demands has unfortunately distracted many adult numeracy practitioners from principles of good assessment practice. (Sadler, 1987, discusses the difference between assessment-led and assessment-driven instruction.) Assessments used for reporting on student achievement should be seen as extensions of the assessments described earlier for diagnostic and formative purposes. In actuality, however, it is a major challenge to design systematic yet simple assessment methods for reporting purposes that represent all key learning outcomes.

A recent major study in Australia designed to address this challenge led to the development of a National Reporting System for adult English language, literacy, and numeracy (Coates, Fitzpatrick, McKenna, & Makin, 1995). This system provides recognition that prior skill repertoire and learning contexts are major factors in considering comparability of results across students and programs. A further study in Australia (McKenna, 1999) has examined the feasibility of establishing a "bank" of clearly delineated exemplars of performance in different general and workplace education contexts as a means of expressing progress on accepted standards or performance continua.

Evaluation

Evaluation of the "success" of a program is not a major focus of this chapter. A few comments are worth making, however. Program evaluation underwent considerable theoretical exposition in the 1970s and early 1980s, whereby the concepts of assessment and evaluation were differentiated. In the 1990s there appears to be a growing convergence, but also lack of clarity, in using the terms, with assessment used not only for documenting student performance but for evaluation of a program's effectiveness as a whole.

The focus of evaluation is not only on what the students gain from their programs but the match between these gains and the intentions and expectations of the program. More broadly, evaluation should identify aspects of a program that were successful, as well as aspects that need to be modified, such as instructional methods, student selection, or curriculum orientation. Padak and Padak (1991, p. 376) provide a simple list of six guidelines for adult literacy program evaluation that are equally applicable to adult numeracy contexts:

- base evaluation on the program's stated goals
- make the evaluation comprehensive
- make the evaluation systematic (not anecdotal)
- use a variety of assessment forms, both qualitative and quantitative
- review the evaluation results in terms of the three categories of program effectiveness: personal factors, programmatic factors, external factors
- use evaluation data to identify parts of the program that need strengthening

Summary. This section focused on the purposes for which assessment in adult numeracy programs might be undertaken, taking as context a broad and proactive definition of numeracy for adult learners. It is hoped that after reading this section, practitioners will examine the purposes for which they are undertaking assessment and adapt assessments to suit different purposes and goals and to reflect the instructional aims and processes of their programs. The following section reviews what constitutes good assessment that can contribute to enhanced and effective student learning.

RETHINKING WHAT COUNTS AS GOOD ASSESSMENT

This section looks at what has traditionally counted as good assessment and presents beliefs about good assessment. In light of the previous discussion, one assessment tool commonly used in the United States, the TABE, is also examined.

Taking a Hard Look at Traditional Assessment

In the past, tests have been designed with the underlying assumption that a test is a method for measuring latent traits or abilities that are not in themselves directly observable but that are measurable by inference. Consider, for example, tests of intelligence. Items on these tests are intended to measure the intangible construct of intelligence by way of specific manifestations or representations of this construct, such as recognition of patterns, richness of vocabulary, and so forth. Likewise, tests of mathematical skills are seen as aiming to examine students' "addition ability"; for example, through their performance on samples of items that presumably require the addition skill to be activated.

The quality of assessment instruments and practices has traditionally been evaluated in terms of their psychometric properties, which are in part determined through statistical studies.

> *Reliability*: the consistency with which a test measures a domain and is free of error. For example, if a test is administered on repeated occasions, for example, six weeks apart with no instruction or expected development occurring in between, the same result should occur. Also, different "judges" evaluating the same open-ended response or performance should, in general, reach similar conclusions.
>
> *Validity*: the degree to which an instrument is representative of the (content of the) domain it is testing. Most reference books talk of four types of validity: face, content, predictive (criterion-related), and construct.

Although tests are usually designed to evaluate the extent to which people possess underlying constructs, the *interpretation of scores* that people obtain on tests is traditionally done through reference to group norms, that is, what others (in a given age/grade level) can do. Such tests make a number of major assumptions such as the distribution of performance in a population, the characteristics of the population itself, the transportability of such performances across contexts and cultures, and more.

The training of teachers (in colleges) has traditionally included courses on topics such as "measurement and evaluation," with an emphasis on psychometric characteristics of tests, mainly reliability and validity (Linn, 1990). It is thus not surprising that many teachers in all education sectors, and certainly bureaucrats or policymakers, turn to standardized tests; such tests appear to satisfy rules of psychometric robustness and are readily available. We need, however, to question whether the principles of reliability and validity are appropriate for adult numeracy assessment. Do we, for example, assume that we are measuring some underlying ability, or are we more interested in assessing observable and actual (or functional) performance, even if it is caused by multiple factors (e.g., multiple math and literacy skills, dispositions, and so forth)?

Principles of Good Assessment

Good assessments have to be both reliable and valid. However, we have to think carefully about what we mean by validity. Messick (e.g., 1989a, 1989b) states

> Validity is an integrated evaluative judgement of the degree to which empirical evidence and theoretical rationales support the *adequacy* and *appropriateness of inferences* and *actions* based on test scores or other modes of assessment . . . hence what is to be validated is not the test or observation device as such but the inferences derived from test scores or other indicator. (Messick, 1989b, p. 5; emphasis in original)

The essence of good assessment, therefore, is that it is appropriate for the context, purpose, and interpretation made or needed. It is also very important to recognize that assessment has to be fully linked with the learning and teaching goals of a program and reflect (for teachers, students, and policymakers) what is valued in a student's performance and learning. A recent report by the Mathematical Sciences Education Board (MSEB, 1994) has enumerated these concerns through three clear principles of good assessment in mathematics:

- The Content Principle: Assessment should reflect the mathematics that is most important for students to learn
- The Learning Principle: Assessment should enhance mathematics learning and support good instructional practices
- The Equity Principle: Assessment should support every student's opportunity to learn important mathematics

In explaining the Content principle, for example, MSEB asserts that assessment tasks that are anchored in important mathematics should "embed mathematics in relevant external situations, require students to communicate clearly their mathematical thinking, and promote facility in solving non-routine problems" (p. 8). These principles are at the core of the *Assessment Standards for School Mathematics*, a key document released by NCTM (1995) that lists additional guidelines and provides many illustrative examples for alternative assessments. Overall, these principles reflect changes occurring in formal mathematics education in schools (Lesh & Lamon, 1992), and they reinforce the often-stated principle that assessment practices need to reflect good instructional theory (Shepherd, 1991).

The need to design better assessments that examine connected knowledge, communication skills, and coping with nonroutine problems, raises many thorny issues that educators and testmakers will grapple with in coming years. While this work continues, it is important to acknowledge that the principles discussed earlier are compelling for adult numeracy instruction, where the starting point for much instruction is the individual's learning needs and life goals. It is critical that assessments used in adult education are capable of delivering to the most important partner in the instructional enterprise, the adult student, complete information on progress and achievement.

Consideration of the TABE

In considering how teachers and students in numeracy classes can identify and work toward development of "important mathematical knowledge," past and present practices of instruction and assessment need to be examined. In this section we examine as a case study how well the Tests of Adult Basic Education (TABE), a key assessment system that is estimated to be in use in over 75% of all adult literacy programs in the United States (Gal & Schuh, 1994), fit with these ideas and principles.

The TABE is published commercially by CTB/McGraw Hill for the purpose of assessing achievement in reading, math, language, and spelling skills. Version 5 of the TABE, which was the test in widest use in the United States when this review was prepared, includes four tests that the test manual describes as assessing overlapping skill levels roughly corresponding to grade levels 2.6-4.9 (Easy), 4.6-6.9 (Medium), 6.6-8.9 (Difficult), and 8.6-12.9 (Advanced). A brief Locator Test is used to determine the approximate skill level of the individual and help decide which of the four tests should be administered to obtain a detailed analysis of the person's skills.

Each of the four Version 5 tests includes seven subtests: vocabulary, reading comprehension, language mechanics, language expression, spelling, mathematical computations, and mathematical concepts and applications. The test is timed and a full administration of all subtests takes almost 3.5 hours. All items in all subtests use a multiple-choice format requiring the test taker to choose one of four possible answers. A person's total score is normally translated into a "grade equivalency" in the range 1.0-12.9, based on conversion tables in the manual.

Before discussing further the content of the TABE, it is important to note that the test manual (Tests of Adult Basic Education, 1987) asserts that it "focuses on basic skills that are required to function in society. Because the tests combine the most useful characteristics of norm-referenced and criterion- referenced tests, they . . . enable teachers and administrators to diagnose, evaluate, and successfully place examinees in adult education programs" (p. 1).

Next is a discussion on the extent to which such mathematical subtests follow the three key assessment principles discussed earlier, and on additional issues pertaining to the meaning of the grade levels determined by the test.

The Mathematical Computation subtest takes 43 minutes and includes 48 multiple-choice items requiring test takers to compute the answer to problems of the type "605 x 256 = ?" or "3.7 - 2.251 = ?" The Mathematical Concepts and Applications subtest takes 37 minutes and includes 40 multiple-choice word problems. Some items briefly describe a situation and ask for a computation; others pose a question about mathematical information presented in a graph, diagram, or coordinate system, usually without any context. (Some examples are analyzed in detail below.)

Contexts. The context within which mathematical problems in the Version 5 TABE are embedded is examined first. In reviewing the TABE, Sticht (1990) noted that only a modest portion of the items are about everyday events in the lives of low-income adults. For example, of the 40 items in the Concepts and Applications subtest in Form M (Medium, grades 4.6-6.9), only one item is about calculating the correct change for a given transaction, no items address savings from bulk purchases, and no items are about the total cost of a purchase with an installment plan and finance charges.

When item contexts in the Concepts and Applications in other TABE tests are examined in a more systematic manner, the deficiencies pointed out by Sticht appear even more glaring. For example, of the 40 items in the subtest in Form D (Difficult), 22 have no context at all (e.g., "which group of numbers contains common multiples of 2 and 9?"). Six items have only a nominal context, usually related to systems of measurement, time, or money (e.g., "How many millimeters are equal to 2 centimeters?"). Seven items have a superficial context (e.g., "It is now 2:30 p.m. What time was it 3 hours and 5 minutes ago?"). Such items appear contrived, as if created for the express purpose of testing mathematical knowledge, not as a simulation of a realistic problem that motivates mathematical reasoning. Only 5 items out of the 40 seem to simulate a real-world problem (e.g., "Mr. Rankin made a deposit of $10 on a hotel room that costs $16 a day. How much money did he owe on the room after 7 days?").

All three assessment principles—content, learning, and equity—appear to be violated by components of the TABE. As new versions of standardized tests are released by test publishers, reviews such as the one provided here (of the TABE Version 5) are bound to become outdated. Yet, new versions of commonly used tests published in the adult education field often retain many of the underlying characteristics of items and tasks used in older versions even when surface aspects of the test are changed. Hence, this review aims to alert readers to some of the features of widely used tests that have to be of concern to test users. Only a few items in the TABE enable test takers to demonstrate real-world mathematical knowledge they may possess. Thus, when used for placement, such tests are likely to underestimate the skills and knowledge that learners possess. On the other hand, overemphasis on school-like tasks means that test takers who have been away from pure computational tasks for years may not perform well on decontextual tasks that require memorization of school-based procedures seldom used in everyday contexts, such as multiplication and division of fractions.

The overemphasis on decontextual problems can also bias the interpretation of scores on pretesting and posttesting in reporting contexts. Some programs are known to use this fact to elevate "gain scores" of students, by administering the TABE to new students *before* the onset of any instruction. By helping students brush up on their general test-taking and procedural skills, programs can demonstrate a sizable gain in TABE scores within just a few weeks. This gain may impress funders, but most of it is artificial and does not represent true change in underlying knowledge or conceptual understanding. Also, score gains may mislead students to believe they have learned more than they really have.

Mathematical content. The TABE Level E (Easy) Computation subtest focuses mainly on whole number operations taught in the first three years of school, with a few items involving fractions and decimals. The Level D (Difficult) Computation Test is almost entirely focused on decimals and fractions, including the notoriously "difficult" division of a fraction by a fraction. The distribution of percent problems in the TABE tests is questionable. Although percents are usually introduced in the school curriculum in grades 5 or 6, there are no items on percent in Test M (Medium), which corresponds to grades 4.6-6.9. Test D (Difficult, grades 6.6-8.9) contains a total of five percent problems, two in the Computation subtest and three in the Concepts subtest. Only one of the later three problems pertains to a realistic context. Furthermore, percentage questions in Form A (Advanced, grades 8.6-12.9) are mostly restricted to the interpretation of percents in pie charts, and do not represent central, real-life situations in which adults have to use or make sense of percentages.

The upshot is that the mathematics being shown as important in these subtests (Content Principle) is very limited. Percent, for example,

is of critical importance in adults' lives; percents are frequently encountered in everyday shopping and financial contexts, and understanding
them is imperative for adults to be considered effective consumers or for
understanding trends and statistics reported in the media (Ginsburg,
Gal, & Schuh, 1995).

Many TABE items also appear to indicate a hierarchy of computational complexity instead of conceptual complexity. With access to calculators that today can handle fractions as well as decimals, are computations necessarily more complex at all? Computational subtests offer little to the adult numeracy student or teacher in terms of the MSEB's
Learning Principle of supporting good instructional practices. The test
can shed little light on students' ability to apply knowledge or on their
problem-solving strategies.

The worst scenario occurs when a test such as the TABE is used
as the entry and departure test for a program and hence becomes the
driving force for the instructional program. The superficial characteristics of computational performance will tend to be addressed within such
a program, in the traditional manner that has already failed so many
adult numeracy students. Indeed, one of the most unfortunate aspects of
testing of this kind is that it *reinforces* the notion for an adult numeracy
student that even on the most complex task, mathematical thinking is
either correct or wrong, a perspective that good numeracy teachers need
to overcome.

Another limitation of the TABE is its strong emphasis on fractions and decimals. Within the mathematics education community, the
jury is still out as to whether fractions and decimals are actually difficult
or just badly taught. Many students attending adult numeracy classes
are likely to have limited mathematics achievement because they have
been exposed to poor mathematics teaching or have left school early. It
is possible that an adult numeracy student could score high on a difficult
computational test by following well-memorized algorithms and procedures, as many schoolchildren did in the past. If a student fails certain
items, the test gives little indication of the source of the problem.
Research such as that of Brown and Burton (1978) has shown that even
in computational areas, people can develop faulty ("buggy") algorithms
that with a little instruction can be remedied.

What underlies performance on the TABE? A focus on strategy,
process, or understanding could be expected to occur in the TABE subtests of Mathematics Concepts and Applications. Again, the multiple-
choice format means that no information is available to the assessor
about the strategies being used by the test taker to solve the questions,
nor possible sources of error and confusion. Good mathematical princi-

ples of application, estimation, and appropriateness of an exact or rough answer are not reinforced. These issues are highlighted here via an analysis of two specific items on the Concepts and Applications subtest in Form D (Difficult).

Example 1: What is another name for 59,600?
F   500 + 90 + 6
G   5000 + 900 + 60
H   50,000 + 9000 + 60
J   50,000 + 9000 + 600

This item represents tasks children complete in primary school to demonstrate their conceptual understanding of place value. It is, of course, possible for a child to learn to complete these tasks without ever having a sense of the purpose of the task nor a concept of place value. It may be possible for adults to forget how they used to"do" these sums because they have developed a keen sense of numerical place value from life experiences, such as in the context of money and commercial transactions. Thus, it makes much more sense to assess knowledge of place value by asking adults (or older children) whether they would rather have $100, $10, $1,500, or $1,050.

Example 2: Which of these can be evenly divided by 3?
F   81
G   83
H   86
J   89

This question is interesting mathematically on a number of grounds. First, an instructor should want to know how a student obtained his or her answer. Ask this question of a few colleagues and find out how they arrived at their answers. If the person was correct, did he or she have mathematical knowledge that incorporated a number of simple heuristics, such as the numerals of numbers divisible by 3 sum to a number also divisible by 3? Did they work out the answer laboriously using a division algorithm and, if incorrect, was it the algorithm that was incorrect or a number fact? Did they have no approach for tackling this sum but in a multiple-choice format were they able to make a guess? Were they allowed to use a calculator and, if so, were they able to use it appropriately to answer this question? Without having answers to these and similar. questions, this item has little value for informing placement decisions.

An important limitation of tests such as the TABE results from their extensive reliance on problems that involve little or no text. Such tests ignore the inherent links between numeracy and literacy skills in everyday functional contexts and do not address the expectation that learners of mathematics are able to demonstrate a facility in "communicating mathematically" (NCTM, 1989; SCANS, 1991). In contrast, the assessment framework used in large-scale literacy surveys such as the National Adult Literacy Survey (NALS) (Kirsch, Jungeblut, Jenkins, & Kolstad, 1993) asserts that "quantitative literacy" is one of several fundamental facets of literacy. Accordingly, the NALS included many functional tasks requiring application of arithmetical operations with numbers embedded in printed materials.

Taken as a whole, the numerous shortcomings of the TABE discussed earlier suggest that, contrary to claims by the test developers, this test is of very limited value as an assessment tool that can support both diagnostic/placement and summative/reporting functions. Nevertheless, a majority of programs in the United States continue to use this test and policymakers continue to require that the TABE (or similar standardized tests) be used for reporting purposes. Such uses ignore the fact that problem-solving strategies, text interpretation abilities, and estimation and number-sense skills do not lend themselves easily to forced-choice assessment formats.

Of course, educators can get additional information from the TABE for diagnostic or formative purposes. Some teachers have created lists of the mathematical content explored by each item, and use such lists to track the number of questions answered correctly in each of many mathematical areas, such as multiplication, division, place value, percents, and so on. This enables teachers to go beyond the "grade equivalency" information and identify specific areas of knowledge where remediation is warranted. Some teachers have also mapped items onto specific sections of textbooks and workbooks and point students to resources they can use on their own. Some teachers have created a diagnostic, qualitative subtest of items from all key mathematical domains covered by the TABE and administer it in an interview format to individual students to study their solution processes and reasoning about problems.

Such steps can alleviate some of the many shortcomings of the TABE. Yet, the lack of context for most items, and the inattention to communicative skills and to reasoning processes during regular administration of the TABE, require that teachers use results from the TABE with great caution and supplement its use by additional measures, some of which are discussed later.

Cultural Differences

One final point of awareness about assessment, particularly with instruments such as the TABE, pertains to the vulnerability of test items to cultural differences. Adults who immigrate to another country (e.g., from Europe to the United States or from Korea to Australia) may be given a written placement test when they first come to an adult education class. Superficial characteristics of the test could have a large impact on the results if teachers do not know much about the nature of students' prior schooling in mathematics and the conventions of the numeration system in the country of origin. Consider the following:

> In Australia, the system of representing numbers can be
> 5,000.00 (old) or 5 000.00 (new)

> In some European cultures this might be written as
> 5 000,00

Will the disappearing (or moving) "," create confusion for an immigrant who is being tested? Are teachers aware of the potential contribution of such phenomena to students' difficulties with reading, communicating about, and computing with numbers?

A broader issue is that of differences in the language used to enumerate quantities and describe mathematical symbols and operations in different countries, and their potential impact on conceptual understanding and learning. The concepts represented by number names in English and Chinese, for example, are quite different, as shown next (phonetic translation of Chinese symbols is given). What could such differences mean for a Chinese person now trying to learn and prepare for work in English?

Number	English	Chinese
1	one	gi
10	ten	shi
100	hundred	bai
1,000	thousand	tsin
10,000	ten thousand	wan
100,000	hundred thousand	shi wan (ten "wan")
1,000,000	million	bai wan (hundred "wan")
10,000,000	ten million	tsin wan (thousand "wan")
100,000,000	hundred million	yi
1,000,000,000	thousand million (Aust.) billion (US)	shi yi

Consider again the second example TABE item presented earlier, "Which of these can be evenly divided by 3"? Would speakers of English as a second language understand the question format and recognize the English language representations of the mathematical concepts "evenly" or "divided by"? For such students, this is a test of language comprehension as much as of mathematical skill. In acquiring another language, the last areas in which fluency is acquired are prepositions and complex verb constructions. Think how the question posed changes if "into" is substituted for "by." Do such tests support the MSEB's principle of equity in either instruction or assessment?

The upshot is that test design and the interpretation of test scores need to be done with caution in many cases. During instruction, teachers could have students discuss the impact of such cultural diversity on assessment and learning, for example, in areas of representations of numbers and number words (as shown earlier), differences in conventions for writing symbols and operations (for denoting long division or multiplication), and culture-specific beliefs about broader aspects of mathematics learning (importance of memorization vs. understanding).

Tapping Into Reasoning Processes

These questions could better meet the principles of good assessment if their format, delivery, and focus were changed, including but not limited to moving from a closed multiple-choice test to an open-ended format in which students construct their own answers and have to explain their reasoning. The solution of even seemingly simple tasks can involve a complex range of strategies and knowledge; it is the identification of these strategies and knowledge that is now regarded as important for effective instructional planning. The amount of information obtained from just a few well-chosen open-format questions can render a lengthy test unnecessary.

Using oral rather than written items, or simulations of real situations rather than word problems describing such situations, can also be very informative, as it may enable adult students to better access their life and work experiences and implicit skills. Lave, Murtagh, and de la Rocha (1984), for example, found that adults who scored an average of 59% on arithmetical problems presented in a written test averaged 98% when confronted by roughly equivalent problems in a real supermarket. Scribner and Stevens (1989) describe a work scenario in a dairy produce factory in which the workers invented and deployed mathematics procedures that were accurate, fast, and flexible, but not based on the standard number base "10" or on heuristics developed in school mathematics. These workers may not know "another name for 59,600," but they were mentally and visually calculating in several number bases simultaneously. Consideration of the real performance of these workers illustrates that teachers need to develop assessment instruments that demonstrate the full breadth of things that students "can do" rather than things they "can't do."

EMERGING PERSPECTIVES ON MATHEMATICS ASSESSMENT

For most teachers today, memories of mathematics assessment at school are of arithmetic computations, geometry proofs, and word problems that are still found in most current textbooks. The emphasis in assessment was on "correctness," although as the problems became more complex, partial marks (scores) could be obtained for being "correct" in some of the reasoning. These types of activities are still valuable for developing skills and logic. However, as assessment activities they all encouraged convergence of mathematical thinking, whereas the essence of problem-solving in real-life situations is divergent thinking. And, as noted earlier, failure to be "correct" discourages students, leading them to consider themselves bad at mathematics (which in those assessments they were).

New trends in assessment in mathematics include tasks and activities that recognize the diversity of mathematical thinking and encourage creativity and divergence of thought. These assessment approaches align with current pedagogical approaches, primarily constructivism (see chaps. by Coben and by Kloosterman, this volume); they emphasize development of understanding and reasoning rather than rote learning and application, and focus as much on process and strategy as on outcome.

One major trend is a move toward *oral and/or group presentations* of authentic problem solving activities. In the United Kingdom this is occurring even at the university level. Such presentations can be very mathematically rigorous, but also emphasize discussion of the logic behind any approach to problem-solving and reflection on the quality and adequacy of strategies employed in the process.

Another major trend, *portfolio assessment*, which is being developed in elementary and secondary education in many countries, including the United States, offers possibilities for adult numeracy assessment. A portfolio parallels the notion of an art student's portfolio, that is, it is a means of collecting examples of different types of work a student has undertaken; it can include examples of best work and/or chart progress through a course, with examples of the student's work at various stages (beginning, weekly, conclusion) of a course. The immediate benefits of the later examples as feedback to students are obvious. Such a portfolio can provide tangible and meaningful evidence of their progress.

Portfolios are gaining in popularity as assessment tools (see, e.g., Broadfoot, 1988), although the concept of "portfolio" is being operationalized differently in different countries. In general, in an adult numeracy setting, learners could document their mathematical knowledge and strategies on entry (using a broad definition of numeracy), record their goals, progressively note goals reached and the increase in mathematical activity, and their final achievements as noted earlier. These final achievements could incorporate standardized tests if appropriate. Even in a short course of three weeks, students could document how much mathematical knowledge they have that they did not know they had, how many mathematical activities they take part in that they did not realize were mathematical, and probably even generic areas of mathematical understanding that they were unaware they had.

These approaches should not be perceived as "soft." They can be highly structured. However, in situations in which information on progress and summative assessment is required for reporting and funding, a number of issues, such as validity and comparability of portfolios, are still being investigated. A likely solution appears to be the development of specific criteria for the grading of portfolios rather than procedures for norm referencing. Systems of peer review of sample portfolios for comparability of standards, known as *moderation*, may also provide checks on reliability and validity.

Perhaps counterpoint to the move in mathematics education toward exploration of process and problem-solving strategy has been a move in vocational education in the United Kingdom, Europe and Australia toward competency-based instruction and assessment. Many countries have also developed lists of core or generic competencies that are considered prerequisite for competent performance in the workforce, not only in vocational education. (In Australia, these core competencies are described in *Investigating Mathematical Ideas and Techniques*; see Mayer, 1992). "Competency-based" relates to the identification of specific tasks in work or life and the successful completion of these. A list of competencies can be extensive and include overarching competencies and layers of subtasks. An example of a numeracy competency could be:

> Can calculate the total of a dinner bill for a three-course meal
> with drinks for two customers
>   - or -
> Can arrange seating for a meeting of 12 people

Although lists of core competencies may be gaining acceptance, this area is not without debate at present, particularly as competencies are expressed in terms of minimum performance and assessed as either achieved or not achieved. Issues of standards of performance and the difference between competence and expertise are being raised. An additional criticism of competency-based assessment in mathematics and numeracy is that it does not allow students to do extended analyses, solve open-ended problems, or display a command of complex relationships (Resnick, 1987), mentioned earlier as a critical element in good instruction and in numeracy assessment. (However, the approach is similar to the identification of job performance characteristics through task analysis and the development of corresponding criterion-referenced assessment activities, which is common in workplace training contexts in which some adult numeracy practitioners function.)

Wolf (1991) has argued that core skill competencies are by definition inseparable from the contexts in which they are developed and displayed, and thus should not be assessed separately. Wolf's position is shared by other recent perspectives on the need for workers to have integrated skills that are more than just an aggregate of individual competencies (see Carnevale, Gainer, & Meltzer, 1990; O'Neil, Allred, & Baker, 1992; Pollak, 1997; SCANS, 1991). Similarly, the Mathematical Sciences Education Board (1994), in discussing the principles that should underlie the development of high-quality assessments, also asserted that,

> Many of the assessments used today, such as standardized multiple-choice tests, have reinforced the view that the mathematics curriculum should be constructed from lists of narrow, isolated skills that can be easily disassembled for appraisal. The new vision . . . requires a curriculum and matching assessment that is both broader and more integrated. (p. 9)

Another major development in assessment has been the focus on *anchored* or *authentic* assessment (Cumming & Maxwell, 1999). These approaches recognize both the differential performances that can occur in assessment divorced from, as opposed to contextualized in, realistic settings, an important difference already noted for adult numeracy students (Lave, Murtagh, & de la Rocha, 1984). They also highlight the complexity of problem solving in contextualized real life and work activities in which first we "have to generate the problems to be solved and then have to find relevant mathematical information" (The Cognition and Technology Group at Vanderbilt, 1990, p. 5). The report by this group provides a good discussion of both anchored and authentic assessment.

SUMMARY AND IMPLICATIONS

Assessment has to be clearly related to and directed by the instructional focus of the class or program and express what is valued regarding what students are to know, do, or believe (Webb, 1992). One key goal of adult education is to enable learners to make sense of the world *outside* the classroom and take action in functional contexts. The field of adult numeracy needs to reconsider its perspectives on assessment given recent changes in the fields of mathematics education and assessment, and assessment theory and practice in general, as well as in light of perspectives on adult numeracy. On the basis of these considerations there are a number of implications for future assessment practice in adult numeracy:

1. Both instruction and assessment of adult numeracy skills should be informed by broad definitions of numeracy and encompass the work and life mathematical experiences and strategies adults already have.
2. Ideally, assessment should address reasoning processes and (mathematical) problem solving, conceptual knowledge and computation, and the ability to interpret and critically react to quantitative and statistical information embedded in print or media messages, as well as examine transfer of mathematical problem solving across life and work contexts.
3. Assessment should be directed by the instructional focus and goals of the program, not vice versa.
4. One type of assessment alone (e.g., use of standardized tests) will not be sufficient to inform all assessment or evaluation requirements of learners or a program.
5. Convenient and apparently simple assessments such as standardized tests may not be appropriate or informative and may do a disservice to students, teachers, and a program.
6. Adult numeracy assessment should encompass the range of assessment forms being used in other educational settings and may include oral reports, group activities, portfolios, and so forth.
7. Adult numeracy assessment should recognize that adult learners may perform at quite different levels in oral mathematical discussions than on written tasks.
8. Assessment indicators for workplace programs are most appropriately drawn from a task analysis of work.
9. Assessment should inform students in a systematic way of their progress in, and achievement from, a program.

10. Only appropriate interpretation and use should be made of assessment information; adult numeracy practitioners need to be aware of cultural differences in planning and interpreting assessment.

Although the above ideas may be appealing, we need to consider that many practitioners in adult education have very limited access to preservice training and professional development opportunities, in part due to funding issues; also, many work part time, at least in the United States, and lack a strong background in math education (Gal & Schuh, 1994). Few resources and little time will be available to train teachers in using more informative, yet also more demanding and costly, alternative assessments. Under such circumstances, teachers and programs may continue to rely heavily on standardized tests, as they are convenient to administer and easy to score, and their results, although not too meaningful, are easily reportable. Funders, in turn, will expect programs to continue to administer such tests, leaving the situation that was lamented earlier largely unchanged.

How can the "poor assessment" cycle be broken? Of course, a change in policies or official reporting schemes would greatly help the process. It has been shown that a top-down process of change has a limited chance for achieving a measurable impact on a complex field such as mathematics education (Lindquist, 1994). A top-down change may not work at all in a majority of adult education programs, those based on local and independent management, on volunteer operations, or that emphasize learner-based approaches. An open, multipronged and long-range approach is needed that assumes that many players with equal stakes are involved in and can affect the change process (Sashkin & Egermeier, 1993).

Many teachers may be presently unaware of the promise of new assessment forms as they have had few chances to try them out in a systematic way and for a reasonable duration of time. University-based researchers and developers can advocate and present relevant information and suggestions, as has been done in this chapter. Yet, it is up to practitioners to begin to experiment with new forms of assessments in collaboration with their students, colleagues, and program administrators, in order to discover what types of improvements in teaching, learning, and achievement can be realized.

Teacher inquiry can take on different forms. Some teachers may decide to informally try one of the ideas presented in this chapter in their classroom; others may create small teams to experiment with and explore one or two ideas within their program and create a supportive environment in which informal yet systematic reflection can develop. Depending on local conditions and motivations, it is desirable to implement a more elaborate and systematic inquiry process in which a group of teachers and administrators from several programs is involved over a period of time; participants can aim to document and publish personal accounts of experimentation and change, as well as summaries of the group effort as a whole. Such practitioner inquiry projects or study circles have been implemented in different parts of the United States (see Lytle, Belzer, & Reumann, 1992, and the chapters by Ariolla and by Curry, this volume, for an illustration of this approach).

Teachers have a vested interest in providing the most effective and valuable instruction for their students. Because implementation of alternative assessments will require more intellectual and time investments from teachers, there is little that can be done by external players in the field to force teachers to change their assessment practices (Ball, 1990). Teachers need to carefully examine and experiment with new assessment procedures in order to better cover the full range of skills and dispositions expected of numerate adults. They will then be prepared to argue for fair and valid uses of assessment for both instructional and accountability purposes and will have access to powerful examples to support their arguments.

REFERENCES

Ball, D. L. (1990). Reflections and deflections of policy: The case of Carol Turner. *Educational Evaluation and Policy Analysis, 12*(3), 263-276.

Bishop. A. (Ed.). (1988). *Mathematics education and culture.* Dordrecht, The Netherlands: Kluwer Academic.

Broadfoot, P. (1988). Profiles and records of achievement: A real alternative. *Educational Psychology, 9,* 291-297.

Brown, P. S., & Burton, R. R. (1978). Diagnostic models for procedural bugs in basic mathematical skills. *Cognitive Science, 2,* 155-192.

Carnevale, A. P., Gainer, L. J., & Meltzer, A. S. (1990). *Workplace basics: The essential skills employers want.* San Francisco: Jossey-Bass.

Coates, S., Fitzpatrick, L., McKenna, A., & Makin, A. (1995). *National reporting system. A mechanism for reporting outcomes of adult English language, literacy and numeracy.* Melbourne: ANTA/DEET.

Cognition and Technology Group at Vanderbilt (1990). Anchored instruction and its relationship to situated cognition. *Educational Researcher, 19*(6), 2-10.

Cumming, J. J. (forthcoming). *Basic mathematical skills and numeracy: A comparison of school-age and adult numeracy students' knowledge and reasoning.*

Cumming, J. J., & Maxwell, G. S. (1999). Contextualising authenic assessment. *Assessment in Education: Principles, Policy and Practice, 6*(2), 177-194.

Gal, I., & Schuh, A. (1994). *Who counts in adult literacy programs? A national survey of numeracy education* (Tech. Rep. No. TR94-09). Philadelphia: University of Pennsylvania, National Center on Adult Literacy.

Ginsburg, L., Gal, I., & Schuh, A. (1995). *Links between informal and formal skills: Adults' knowledge of percents* (Tech. Rep. TR95-17). Philadelphia: National Center on Adult Literacy, University of Pennsylvania.

Kirsch, I. S., Jungeblut, A., Jenkins, L., & Kolstad, A. (1993). *Adult literacy in America*. Princeton, NJ: Educational Testing Service, National Center for Education Statistics.

Lave, J., Murtagh, M., & de la Rocha, O. (1984). The dialectic of arithmetic in grocery shopping. In B. Rogoff & J. Lave (Eds.), *Everyday cognition: Its development in social contest* (pp. 67-93). Cambridge: Harvard University Press.

Lesh, R., & Lamon, S. J. (Eds.). (1992). *Assessment of authentic performance in school mathematics*. Washington, DC: American Association for the Advancement of Science.

Lindquist, M. M. (1994). Lessons learned? In I. Gal & M. J. Schmitt (Eds.), *Proceeding—1994 Conference on Adult Mathematical Literacy* (pp. 91-97). Philadelphia: University of Pennsylvania, National Center on Adult Literacy.

Linn, R. L. (1990). Essentials of student assessment: From accountability to instructional aid. *Teachers College Record, 91*(3), 422-436.

Lytle, S. L., Belzer, A., & Reumann, R. (1992). *Invitations to inquiry: Rethinking staff development in adult literacy education* (Tech. Rep. No. TR92-2). Philadelphia: National Center on Adult Literacy, University of Pennsylvania.

Mathematical Sciences Education Board (MSEB). (1994). *Measuring what counts*. Washington, DC: Author.

Mayer, E. (Chair). (1992). *Putting general education to work: The Key Competencies Report*. Australia: The Australian Education Council for Vocational Education, Employment and Training.

McKenna, R. (1999). *Assessment materials development for literacy and numeracy identification and outcomes reporting in VET*. Melbourne: COMMET.

Messick, S. (1989a). Validity. In R. L. Linn (Ed.), *Educational measurement*. New York: American Council on Education and Macmillan.

Messick, S. (1989b). Meaning and values in test validation: The science and ethics of assessment. *Educational Researcher, 18*(3), 5-11.

Mikulecky, L., & d'Adamo-Weinstein, L. (1991). Evaluating workplace literacy programs. In M. C. Taylor, G. R. Lewe, & J. A. Draper (Eds.), *Basic skills for the workplace* (pp. 485-499). Toronto: Culture Concepts.

National Council of Teachers of Mathematics (NCTM). (1989). *Curriculum and evaluation standards for school mathematics.* Reston, VA: Author.

National Council of Teachers of Mathematics (NCTM). (1995). *Assessment standards for school mathematics.* Reston, VA: Author.

O'Neil, H. F., Allred, K., & Baker, E. L. (1992). *Measurement of workforce readiness competencies: Review of theoretical frameworks.* Los Angeles: University of California at Los Angeles, National Center for Research on Evaluation, Standards, and Student Testing.

Padak, N. D., & Padak, G. M. (1991). What works: Adult literacy program evaluation. *Journal of Reading, 34,* 374-379.

Perkins, D. N., & Salomon, G. (1989). Are cognitive skills context bound? *Educational Researcher, 11*(6), 639-663.

Pollak, H. O. (1997). Solving problems in the real world. In L. A. Steen (Ed.), *Why numbers count: Quantitative literacy for tomorrow's America* (pp. 91-105). New York: The College Board.

Resnick, L. B. (1987). *Education and learning to think.* Washington, DC: National Academy of Sciences.

Sadler, D. R. (1987). Specifying and promulgating achievement standards. *Oxford Review of Education, 13,* 191-209.

Sashkin, M., & Egermeier, J. (1993). *School change models and processes: A review and synthesis of research and practice.* Washington, DC: U.S. Department of Education, Office of Educational Research and Improvement.

Scribner, S., & Stevens, J. (1989). *Experimental studies on the relationship of school math and work math* (Tech. paper #4). New York: Teachers College, Columbia University. National Center on Education and Employment.

Secretary of Labor's Commission on Achieving Necessary Skills (SCANS). (1991). *What work requires of schools: A SCANS report for America 2000.* Washington, DC: U.S. Government Printing Office.

Shepherd, L. A. (1991). Psychometricians' beliefs about learning. *Educational Researcher, 20*(7), 2-16.

Sticht, T. G. (1990). *Testing and assessment in Adult Basic Education and English as a Second Language programs.* San Diego, CA: Applied Behavioral and Cognitive Sciences. (Available through the U.S. Department of Education, Division of Adult Education and Literacy, Washington, DC).

Sticht, T. G., & McDonald, B. A. (1993). *Automotive trades information processing skills: Mathematics.* Westerville, OH: Glencoe.

Taylor, M. C., Lewe, G. R., & Draper, J. A. (Eds.). (1987). *Basic skills for the workplace*. Toronto, Canada: Culture Concepts.

Tests of Adult Basic Education (1987). *Examiner's manual*. Monterey, CA: CTB/McGraw Hill.

Webb, N. L. (1992) Assessment of students' knowledge of mathematics: Steps towards a theory. In D. A. Grouws (Ed.), *Handbook for research on mathematics teaching and learning* (pp. 661-686). New York: Macmillan.

Venezky, G. L., Wagner, D. A., & Ciliberti, B. S. (Eds.). (1990). *Towards defining literacy*. Newark, DE: International Reading Association.

Venezky, R. L. (1992). *Matching literacy testing with social policy: What are the alternatives?* (Policy Brief No. PB92-1). Philadelphia: University of Pennsylvania, National Center on Adult Literacy.

Willis, S. (Ed.). (1990). *Being numerate: What counts?* Hawthorn, Victoria: Australian Council for Educational Research.

Wolf, A. (1991). Assessing core skills: Wisdom or wild goose chase? *Cambridge Journal of Education, 21*(2), 189-201.

16

Assessment of Adult Students' Mathematical Strategies

Mieke van Groenestijn
**University of Professional Education
Utrecht, Netherlands**

Adult Basic Education (ABE) is often the last opportunity for adults to learn what they missed in former days, as they try to improve their chances in community and work life. This chapter focuses on the challenges involved in assessing mathematical reasoning strategies that students possess when they first come to an ABE class. It describes an assessment instrument, The Supermarket Strategy, that was developed in The Netherlands to evaluate mathematical strategies of students and discusses the training necessary for using this instrument. Implications for the way in which assessment should be carried out to reflect recent reform efforts in mathematics education are explored.

ABE AND MATHEMATICS EDUCATION IN THE NETHERLANDS

Adult education has received little public attention and funding in the Netherlands until the late 1970s. The influx of immigrants from former

Dutch colonies and "Mediterranean" countries who came to work in Holland, and the realization that more students than previously thought had been finishing regular Dutch schools with inadequate skills, caused the government to establish a national system of educational institutes for helping adults improve their skills.

While a new system for adult education was being put in place in the early 1980s, mathematics education at all levels in The Netherlands was radically changing, and the reform was termed Realistic Mathematics Education (RME) by educators.[1] RME is based on the work started by Hans Freudenthal, who in 1970 founded the Institute for the Development of Mathematics Education, the current Freudenthal Institute, with the aim of changing the traditional computational approach to mathematics education that was prevalent in Holland and elsewhere until the 1970s. A 10-year research and development effort at this Institute led to a program (Freudenthal, 1983; Gravemeijer, 1994; van den Heuvel-Panhuizen, 1996; van den Heuvel-Panhuizen & Gravemeijer, 1991; Streefland, 1989; Streefland, 1993) that was adopted by a large majority of schools in the Netherlands in the early 1980s, as well as by much of the adult education community.

RME emphasizes the use of realistic context problems, representations of reality and models to relate classroom instruction and learning to the student's real environment and real experience. It aims to practice skills and enhance students' ability to apply knowledge and skills in practical situations. RME emphasizes the development of individual reasoning processes and problem-solving strategies. It advocates allocation of a lot of classroom time for interaction in groups (assuming learning is a product of social interaction as much as of individual processes) and promotes the interrelation of subject areas.

Formerly, math education in The Netherlands (and in many other places as well) consisted for the most part of learning calculations. The answers to problems were either right or wrong, so the teacher assessed math knowledge by checking answers and adding up a score. Now, math education involves much more: students learn to solve realistic math problems. They start by analyzing a mathematical situation, which is usually a problem in a real-life context (not merely a "word

---

[1]The RME approach was already operating in many Dutch schools before work started in the United States on what later became known as the *NCTM Standards*, described in chap. 3 and elsewhere in this volume. The NCTM Standards share some key ideas with the RME approach. A middle-school curriculum based on RME was under development for several years by a joint American-Dutch team working at the National Center for Research in Mathematical Sciences Education at the University of Wisconsin-Madison, and was recently made available to U.S. educators. It contains several modules and many ideas that would be of interest to adult educators and their students.

problem"), and translate it into math formulas that can be used to solve the problem. They learn to rely on their own ways of problem solving and to review their own solutions.

This process can be better understood by examining some real-life problem situations. For example, a person is buying cookies (or cupcakes, or balloons) for a children's party. There must be enough cookies for all the children. Determining how many packages to buy depends on how many cookies are in a package. If there are 8 cookies in a package, how does one calculate how many packages will be sufficient for 40 children? How would a person solve a similar problem if he or she wants enough for a school of 384 pupils and there are 12 cookies in a package? Think about the way one would solve the problem with a colleague or with students. What kind of math skills and reasoning strategies are needed to be able to solve these problems? Or, study the next example: A person gets 4% interest a year on his or her bank savings account. How much interest will she get on $500 after one year? What will they get if they cancel their account after 7 months? What mathematical knowledge and skills are involved in this second question?

Within the context of establishing a new system for adult education and adopting the RME model of instruction, adult educators in the Netherlands needed to be able to use assessment tools that would allow them to analyze adults' mathematical strategies and knowledge, help discover where students must begin their studies, and also enable them to monitor students' progress and evaluate the quality of ABE instruction. However, testing adults in ABE had no history; no general standardized math test had been developed in Holland for low-educated and illiterate adults. The most common tests were developed for children, teenagers, or adolescents in a school situation. There was a need to develop a new approach to the assessment of mathematical skills of ABE students that would be based on RME principles.

CONSIDERATIONS IN DESIGNING AN RME-BASED ASSESSMENT TOOL

The RME model described earlier implies that assessment instruments must consist of functional problems in an everyday-life context that can be solved in different ways. The student must be free to use his or her own methods. This model assumes that assessment cannot be focused only on "product" and must be focused more than before on "process." A good answer is important, but the process by which the student solves the problem gives more information about his or her way of thinking and the quality of his or her calculations. The following questions become important in this context:

- Does the student understand the context?
- Is he or she able to translate the context into mathematical formulas?
- Which formulas does he or she discover in the context?
- Are the formulas appropriate for the context?
- How does he or she calculate?
- Is the answer correct?
- Does the student check his or her own answer?
- Is the student able to use previous calculations in a new situation when this is possible?

In addition to these general points,, in order to be useful in planning instruction for new ABE students, the structure, content, and administration of an RME-based assessment instrument should also (a) take into account the characteristics of ABE students in the Netherlands, (b) enable systematic exploration of students' familiarity and mastery of all key topics or subareas of mathematics deemed relevant for ABE students, and (c) consider different levels of difficulty of functional tasks (e.g., the interest example just given is at a much greater level of difficulty than the cookies problem). Next I discuss these design principles. In the remainder of the chapter I describe in more detail the nature of an assessment instrument, the "Supermarket Strategy," that was developed based on these principles, and discuss implications for needed professional education of ABE teachers (van Groenestijn, Matthijsse, & van Amersfoort, 1992).

Nature of ABE Students Learning Math in the Netherlands

The majority of students in ABE in the Netherlands come from other countries and have little or no school experience. Often they are not used to learning in a "scientific way," for example by reading a book or working on written exercises. Yet, while they have not developed many academic skills, they have learned much through practical experiences and by thinking and talking about those experiences. Their repertoire of strategies increases by trial and error and by exchanging their experiences with family, neighbors, friends, colleagues, and so on. They remember those strategies that they really understand and that are easy to use. Also their math knowledge and skills have developed in this way. Because they learn in this manner, their knowledge and skills are often context-bound and fragmented, but at the same time quite practical and functional.

Teachers in ABE need to learn what their new students already know about mathematical issues and how they apply their mathematical knowledge in everyday life. Their everyday math strategies should be starting points from which learning connects and builds. Good strategies should be stimulated to become more generalizable. Incorrect strategies must be transformed into good ones. New, more effective strategies can be learned.

In ABE teachers also work with adults who failed at school when they were young. They often have negative memories of school. Therefore, testing can be a frightening experience for them. Illiterate adults are not accustomed to testing situations at all and cannot read a test. Moreover, there are many students who are not native speakers of the language. It is often difficult for them to speak, read, and write in Dutch, so a written test can be problematic. Thus, a mathematics assessment instrument that is not threatening, is geared to adults, and does not depend much on the ability to read is needed. Oral forms of assessment are optimal, although if students are already able to perform calculations on paper, it may be possible to give them a written test as well.

Classifying Skill Area and Skill Level

In planning the assessment of students' knowledge and skills, the skill areas to be assessed and the levels of mastery or knowledge to be used in classifying students' responses must be decided first, so that meaningful instructional decisions can be made. The results of an analysis of the mathematical skills and knowledge an adult needs for optimal functioning in The Netherlands that took place during 1988-89 formed the groundwork of the decisions. This analysis suggested three basic areas related to mathematical knowledge:

1. *Basic Skills*: counting and numbers, addition and substraction, multiplication and division
2. *Proportions*: proportions, percents, fractions, and decimals
3. *Measurement and Geometry*: constructing in space, the metric system, money calculating, time, and the calendar.

After an extensive study, the three areas just listed were subdivided into seven fields of knowledge, and performance goals were formulated for six levels of mastery/knowledge within each field, as depicted in Figure 16.1.[2]

---

[2]Because calculating with money in the Netherlands is like calculating with decimals in the metric system, educators are faced with an "instructional trap." Many adults are able to calculate with money, so it seems they have mastered the metric system. But then they are not able to calculate with meters and kilometers or with measures of weight (e.g., kilograms) and volume (e.g., liters). Calculating with money was thus considered in this scheme as a separate, important functional field.

		Elementary levels		Intermediate levels		Advanced levels	
Mathematical Areas	Fields	1	2	3	4	5	6
A. *Basic Skills*	1. counting and numbers; addition and subtraction						
	2. multiplication and division						
B. *Proportions*	3. proportions and percents						
	4. fractions and decimals						
C. *Measurement and Geometry*	5. Metric system and geometry						
	6. money calculating						
	7. clock and calendar						

Figure 16.1. Skill grid

In this scheme there is a horizontal and a vertical structure. Horizontally, the subject matter per field is built up in six levels of complexity or difficulty, from elementary to advanced. Vertically, there is a coherence between the three main areas (which are divided into seven fields). A specific topic such as "doubling and halving," for example, will appear in level 2 and serve as the starting point for addition, subtraction, multiplication, and division; it is also the starting point for working with proportions and the metric system at the same level. At level 3 "doubling and halving" is done twice: it is helpful for coaching the students in learning times tables, and it is also the starting point for working with fractions, decimals, and percents. Students at level 3 must be able to make optimal use of multiplication and division. Additionally, they must have insight into the use of fractions, decimals, and the metric system up to a half, a quarter, and a tenth of something like weight, volume, or length.

As the classification system of student skills was developed, it became necessary to evaluate the functionality of some mathematical concepts and applications in contemporary life and work situations. For example, the developers asked: What do adults need to know about fractions for optimal functioning in daily life? Technological developments such as computers and calculators have influenced the application of problem-solving strategies. The combination of estimation, mental calculation, and the use of a calculator is much more relevant to daily life than calculating with algorithms on paper. In the assessment process, adult learners should thus be encouraged to use such strategies rather than follow set procedures and use algorithms.

Use of the calculator requires knowledge and application of decimals. This technological development makes calculating with decimals much more important in our society than manipulating fractions. Therefore, the "Supermarket Strategy" was designed so that competency with fractions could be demonstrated by a conceptual understanding, the use of appropriate fraction language, and the ability to perform some basic operations with benchmark fractions.

The classification of difficulty (horizontal axis in Figure 16.1) is characterized by:

- An advance in computational skills from an elementary level up to more advanced, multistep procedures in each field.
- A gradual transition from more concrete contexts at the lower levels to more abstract contexts at the higher levels. For students at elementary level, real objects such as money and a measuring cup may be used to solve given problems. At the intermediate level, the contexts are mostly represented by pictures and by symbols and numbers. At the advanced level, the contexts are represented more abstractly through the use of symbols and supported by pictures. Note that representations are important at all levels.
- A progression from simple contextual problems at the lower levels to more complex ones at the higher levels.

Knowing the learner's mathematics level makes it possible to plan an appropriate curriculum and to place students in elementary, intermediate, and advanced groups. It is important to remember that, even when appropriately placed, all learners will still need their own individual guidance. In addition to allowing educators to follow each student's progress during the year, the results of the assessment can also be used to develop course programs and textbooks in ABE that are based on this level classification. (Should readers from other countries want to use such a level classification, they will need to adapt it to their country because of differences in money, measurement, and other systems.)

THE "SUPERMARKET STRATEGY"

The "Supermarket Strategy" refers to the decision to design an assessment instrument based on everyday life problems encountered in the context of supermarket shopping; this context is recognizable for virtually every adult in The Netherlands (including newcomers because that even they shop in supermarkets). The use of "strategy" highlights our interest in exploring students' thinking strategies. It also suggests that the instrument provides teachers with a strategy for figuring out the student's capabilities; this strategy is based on collecting and evaluating qualitative information in accordance with the classification scheme depicted in Figure 16.1, using an *adaptive* procedure. The result of the process is a *profile* of the students' capabilities that combines both qualitative and quantitative elements, rather than a single summary "standardized score" or "grade level" as often used in math tests.

This section outlines the main elements of the assessment, including the nature of the problems and stimuli presented, and the approach to interviewing and observing students. An interview excerpt illustrates some of the issues that may come up during an actual assessment.

CONTENT OF PROBLEMS

In the Supermarket Strategy a series of 60 context-rich problems was developed that represented all the cells created by crossing skill fields with levels of difficulty as shown in Figure 16.1. Some cells have more than one problem. Despite the name of our assessment method, not all problems relate only to purchasing situations. In the field of measurement, for example, a distinction is made between length and area, volume, and weight. In the field of "clock and calendar" there are separate questions about clock and calendar problems. At the intermediate and advanced levels there are other contexts. Yet, whatever context is presented, and whether it involves direct computation or other mathematical processes, students are always free to approach a problem in their own way, using their own strategies.

Each problem is described on a separate "Context Card" and relates to an everyday situation in which adults have to apply counting, adding, subtracting, multiplying, dividing, and so on. The adult is free to create his or her own strategy for solving the presented problem, but each context requires knowledge and the application of key mathematical concepts related to the cells in Figure 16.1.

A special advertising leaflet (flyer) has been developed, in collaboration with a large supermarket chain in The Netherlands, to provide a realistic visual and informational context for the supermarket-related problems. The leaflet presents, just like the real thing, many pictures of different kinds of products, with accompanying prices. The products are grouped and displayed in such a way that a discussion can focus on many issues, starting from simple issues such as the shapes and relative sizes or volumes of different containers (e.g., milk cartons, bottles) or the number of elements in different pictures (e.g., bottles in boxes), and culminating with advanced issues, such as estimating the total price of a mix of products.

The use of a "natural" leaflet enables the examiner and examinee to talk together about several topics in a natural way that resembles an adult conversation and does not appear to be like a traditional test. This conversational approach, which is based on observations and interviewing, can provide the examiner with information about the student's problem-solving strategies, while allowing the student to reflect on them as well.

Interview and Observation Methods

The primary methods of obtaining information in the Supermarket Strategy are interviews and observations. Using the leaflet and context cards, the teacher conducts an interview with the student, based on three types of questions:

1. How are you going to do it?
2. How are you calculating/what are you doing now?
3. How did you do it?

The first question refers to how the student will manage the math problem; as explained later, this question is presented *before* students start working on a problem, and only to those who function above the elementary levels. The second question is meant to discover which problem-solving strategies the student uses *during* his or her solving or calculation. The third is presented *after* the student provides an answer and enables the interviewer to discuss the strategy used and to reflect with the solver on what was done and why. The three questions together enable the interviewer to follow the problem-solving procedure very closely. In this way a *qualitative* analysis can be made of the student's mathematical skills and knowledge.

Encouraging the student to talk out loud about the way he or she is solving the problem is very important. As long as the student is talking, his or her thinking process is directly clear to the interviewer. Sometimes the student does not say a word during calculating or answers quickly without performing any written calculations. In that case the teacher must ask the student to give feedback on the method of his or her calculation. Sometimes it is clear that the student did not know what he or she was doing and was just following a calculation procedure he or she learned to do previously. At other times the student may have used a strategy he or she mastered before but cannot verbalize.

It may also be possible that the student has not mastered the language. In this case the teacher continues to ask questions, using visual representations of the problem or a technique that called "reflecting" on the math problem. In reflecting the teacher repeats what he or she saw the student do and then asks the student if the observation was correct. The student needs to answer with yes or no, and provide clarification if needed. Visual representations and "reflecting" are very important techniques to use in talking with non-native speakers.

With higher functioning students who are working on more complicated problem contexts, it becomes necessary to ask the student how he or she will manage the problem *before* he or she start to solve it. Talking about the problem-solving process before actually doing it is an implementation of the RME method. It is much more difficult than talking aloud during the problem solving or reviewing after it.

Sometimes during the problem-solving process the student may get stuck. Then the teacher is allowed to help the student in certain ways, such as:

- The interviewer may structure the task. For example, difficult numbers can be split up into simpler numbers: $357 = 300 + 50 + 7$
- The teacher may give the student a good strategy. For example, with the problem $22 + 23$, he or she may ask, "Do you know what you get by doubling 22"? or, ". . . by doubling 20?"
- It is possible to give the student a similar but easier task. For example, for the problem $52 - 6$, the teacher can ask, "How much is $50 - 6$ or $12 - 6$?"
- The teacher may give the student some concrete materials, such as money or a measure cup.

- The teacher may try calculating together with the student to get to a solution, especially if the student seems to give up.
- The teacher may call the student's attention to a mistake in his or her calculation.
- In a progress test situation the teacher may refer to a lesson situation in which a similar problem has been taught.

While helping in this way, the teacher can collect qualitative information about the student's solving procedures, math ability, math knowledge, and numeracy skills. These can then serve as basic information for helping the student with learning mathematics.

The assessment result will partially depend on the expertise of the teacher. The better trained the teacher is, the faster he or she will get information and the information will be of a higher quality. A well-trained teacher should be able to determine the student's math level and skills in about 30 to 45 minutes.

The Assessment Procedure

The Supermarket Strategy can be used for intake assessment, to assess progress in the class setting, and as a final assessment as the student leaves ABE. In an intake situation the examiner/teacher determines the student's skill profile using an adaptive process, based on only a subset of the 60 context cards; in other words, not all problems will be presented to a student. The examiner starts at a difficulty level (usually 1, 2 or 3) that seems most suitable for the examinee, based on a "locator" problem. Assessment stops when it appears the examinee is unable to solve more problems in a reasonable manner.

An example for a locator task is the "bread" problem, which helps select which of the four skill levels to start with. The leaflet includes pictures of different kinds of breads and rolls. The teacher may ask the student, for example, to buy two different types of bread. What does he or she have to pay for them? If the student answers well, the next question may become more complex (e.g., "You now want to buy 6 rolls; what will you have to pay for them?"). As part of this process, the teacher analyses the student's ways of adding and multiplying.

The placement decision called for in an intake process does not require that all skills be assessed in an equal manner. Often, only the first three or four fields will be assessed for an initial placement. Despite the order of fields in Figure 16.1, Field 6 (money calculating) is always the first one to be actually used after the locator problem because most people are familiar with money and it is very important in daily life.

Figure 16.2 shows the actual order of administration of problems to a hypothetical student. The path followed by the examiner through the grid is determined by the quality of the student's responses. For each response the student can get 0, 1, or 2 points, which indicate:

0   The student does not understand the context and is not able to solve the problem or may come to a solution in a very slow and laborious way.

1   The student is able to solve the problem, but not quickly and perhaps with many mistakes and restarts. The student is not certain about his or her solution.

2   The student gives a good solution in a smart, fast way, is confident in his or her solution, and is able to reflect on it.

If the student gets a 2, the next question will be in the same field but at the next higher level. If the student gets a 1, the next question will be at the same level but in the field below. If the student gets a 0, the next question will be in the same field but at the next lower level. At the end of the assessment process, a qualitative profile of the student becomes evident. A separate list with a description of possible problem-solving methods can be added.

After a couple of months, when the students are accustomed to Realistic Math Education and the teacher knows more about their reading and writing performance, some context cards at the level corresponding to the course can be given in a lesson situation. The teacher can ask the students to write down their own problem-solving strategies, and after that the students can discuss the different ways of problem solving in small groups. In this way an experienced teacher can assess the students' skills within the class and monitor their progress.

Sequence	level 1	level 2	level 3	level 4	level 5	level 6
locator	2	2	1			
field 6			2	1		
field 1			1	0		
field 2			1			
field 3		1	0			
field 4		1				
field 5						
field 7						

Figure 16.2. Skill grid used in actual assessment showing the qualitative profile of a virtual student

If this manner of assessment is followed throughout ABE instruction, it will not take much time to assess the level and math skills of the student during the course and at its conclusion when the student leaves the ABE phase of his or her education.

A Sample Assessment Interview

The following excerpt is taken from an assessment interview with Farid, an ABE student from Morocco who had some previous schooling. It illustrates how the teacher's help allows Farid to move beyond his fragmented math memories to using his common sense math. The following problem was discussed (see Figure 16.3):

Each box contains eighteen (18) packages of caramel cookies.

How many packages in total?

Figure 16.3.

Farid stares at the picture. He writes 18 and 11 in a vertical notation and draws a line. For a while nothing happens. Then he says, "One times eight equals eight, and one times one equals one." Now 18 is written below the line. For a while, again, nothing happens. He repeats the same story and writes 18 underneath the other 18. He draws another line below and adds: 36. He shakes his head, "It's not correct."

He makes a second attempt. Again he writes 18 and 11 in vertical notation with a line below it. Now he does one times eight equals eight and one times one equals one, but he writes the one below the eight. He repeats this. Now 88 and 11 are written in vertical notation.

He adds, gets 99, and smiles, "This is better."

"Why is this better?", I ask.

"It is more", is his answer.

Yes, it is more, but what is happening here? This young man has had six years of elementary education. Again he writes 11 and 18 in vertical notation. This time 11 is on top and 18 is below it, and he tries again. Eighteen appears below the line and below that a one. He gives up.

"I don't know."

"Would you like me to give you a hand?", I offer.

"Wait. . . ." He makes another attempt and remembers that in the second line the number has to be moved over one spot. He smiles, "Yes, this is right." (He adds a period in the empty spot from which a number was moved one column to the left.)

"Do you think this is difficult?", I ask.

"No, just a lot of thinking", he answers. "Forgot somewhat."

"You did learn this in school?"

"Yes, it's long ago, I forgot somewhat."

"Could you do it mentally as well?", I ask.

He laughs. "No, a lot of thinking, thinking, keep thinking."

Still I try, "If you'd take this box from the pile, how many boxes are left?" (I put my hand on the box on top.)

"Ten boxes", he answers.

"Yes, how many small boxes with cookies do you have then?" "180, that's easy", and immediately, "and 18 equals 198." He laughs, "Yes, that's easy."

Of course, now, I press on. "How did you solve that problem on paper? Why do you write a period in this calculation and put 18 in front of it?" (I point to the last one he did.)

"Yes, that's how you do it."

We do it again, mentally. First he finds that 10 times 18 equals 180 and then adds 18. In the meantime I write 180 and 18 in vertical notation, and we add. He can do this. After that, we look at his last calculation on paper and we compare both.

"That's also one times 18 and 10 times 18."

Then I ask him, while pointing at the period, "Could you also write here a zero?" "No, not allowed, has to be a period," he answers.

TRAINING TEACHERS TO LEAD ASSESSMENT INTERVIEWS

To be able to effectively and responsibly assess students using this method, it is necessary for teachers to be professionally educated in Realistic Mathematics Education and specifically trained in using the Supermarket Strategy.

During the course of their professional education, teachers become confident with RME. For most teachers this is their first exposure to RME because their own math education took place before the reforms. Before training, most teachers are convinced that their own problem-solving strategies are the only good ones. They do not realize that there often are several ways to solve a problem. They have to learn that other people's problem-solving strategies can differ from their own. In addition, they have to learn how math education for adults differs from math education for children. It is important for them to learn what RME really involves and how they can educate ABE students this way.

Specific training in the use of the Supermarket Strategy is needed because the teacher must know the contents of the assessment tool, such as the level classification and the context cards. The teacher must be able to distinguish between effective and less effective problem-solving strategies and understand that it is acceptable for the student to be creative in solving problems. The teacher must be able to lead a good assessment interview and to help the student during the assessment situation, and he or she needs to be considerate in this process. The teacher must also feel confident to make decisions about following all the routes in the level classification during the interview.

The training provided to teachers on using the Supermarket Strategy consists of six sessions of about 3.5 hours each. It includes two real interviews with actual students, in which teachers are expected to lead an interview using good interview techniques, as well as determine the student's math level with the use of the level classification. Video and audiotaping has proved very helpful in this training, and time is also provided for teachers to discuss and reflect on their experiences with colleagues.

IMPLICATIONS AND A LOOK INTO THE FUTURE

The way of assessing adults described in this chapter is qualitative, realistic, and performed in an adult way. Experiences in Holland have had good results and led to some positive side effects. By interviewing the students during the intake and in the lesson situations, the teacher ends up knowing much more about their ways of thinking. That enables the teacher to adjust the lessons to fit the students' needs. For ABE teachers this assessment framework becomes a kind of manual for math education. Assessment through interviewing has helped to change teachers' instructional methods. Their style of instruction becomes more of a style of interviewing and helps to stimulate the problem-solving strategies of the students. Students in ABE tell us that they feel very comfortable with this way of assessment, and, after a period of getting used to this new style of education, they become more confident in their own problem-solving strategies.

Overall, we have found that well-trained teachers using appropriate adult assessment instruments such as the Supermarket Strategy are the starting points for quality in math education in ABE. That said, new changes in adult education in Holland are affecting the role or extent of use of this approach. Recently, in part because of the need to reduce unemployment, Adult Basic Education, which, like all types of education in Holland is financed by the government, was merged into a much broader system of Adult Education that also includes Further Education and Vocational Education. All activities are organized via large institutes and delivered through flexible modules that have to be cost-effective and based on proof of effectiveness. The idea is to train people for jobs in a short time. Assessment procedures and individual tutoring are very important, but were changed to fit this larger system. Basic education programs aimed at low educated and illiterate adults have almost ceased to exist as a separate area of activity, at least for now, and training for ABE teachers has been drastically reduced.

In this context, a new qualification (assessment) scheme is being implemented. Since 1995, a new math test, developed by the Central Institute of Test Development (in Dutch, CITO) is being widely used. It is a written test and yields a standardized score and a grade level. (Does this sound familiar to adult educators from other countries?) The test appears to have various shortcomings and relies quite heavily on language use, so is not too well-suited for non-native speakers. As a result, although usage of the Supermarket Strategy was reduced after the reorganization of adult education came into effect and advocated the use of the CITO math test, adult educators in Holland are now beginning to return to using the Supermarket Strategy due to its ability to provide rich qualitative and quantitative information. Attempts are now being made to update this scheme to simplify its administration and reduce the amount of needed teacher training, with the hope that it will be available to teachers to supplement any standardized test they may have to use.

REFERENCES

Freudenthal, H. (1983). *Didactical phenomenology of mathematical structures*. Dordrecht: Kluwer.

Gravemeijer, K. P. E. (1994). *Developing realistic mathematics education*. Utrecht: CB-β Press.

Streefland, L. (1989). Realistic mathematics education (RME): What does it mean? In C. A. Maher, G. A. Goldin, & R. B. Davis (Eds.), *Proceedings of the eleventh Psychology of Mathematics Education—North American Chapter conference* (Vol. II, pp. 121-124). New Brunswick, NJ: Rutgers University.

Streefland, L. (Ed.). (1993). *The legacy of Hans Freudenthal.* Dordrecht, Boston, London: Kluwer.

van den Heuvel-Panhuizen, M. (1996). *Assessment and realistic mathematics education.* Utrecht: CD-β Press.

van den Heuvel-Panhuizen, M., Gravemeijer, K. P. E. (1991). Tests are not all that bad: An attempt to change the appearance of written tests in mathematics instruction at the primary school level. In L. Streefland (Ed.), *Realistic mathematics education in primary school.* Utrecht, The Netherlands: CD-β Press.

van Groenestijn, M., Matthijsse, W., & van Amersfoort, J. (1992). *Supermarktstrategie, een procedure voor niveaubepalen bij rekenen in de basiseducatie.* Utrecht: Stichting IDEE.

Author Index

Subject Index

Printed in the United States
216799BV00004B/1/P

9 781572 732339